The Economics of Water and Waste in Three African Capitals

RICHARD C. PORTER
LOUIS BOAKYE-YIADOM JR.
ALBERT MAFUSIRE and
B. OUPA TSHEKO

Routledge
Taylor & Francis Group

LONDON AND NEW YORK

First published 1997 by Ashgate Publishing

Reissued 2018 by Routledge
2 Park Square, Milton Park, Abingdon, Oxon, OX14 4RN
52 Vanderbilt Avenue, New York, NY 10017

Routledge is an imprint of the Taylor & Francis Group, an informa business

Notice:
Product or corporate names may be trademarks or registered trademarks, and are used only for identification and explanation without intent to infringe.

Publisher's Note
The publisher has gone to great lengths to ensure the quality of this reprint but points out that some imperfections in the original copies may be apparent.

Disclaimer
The publisher has made every effort to trace copyright holders and welcomes correspondence from those they have been unable to contact.

A Library of Congress record exists under LC control number: 97070347

ISBN 13: 978-1-138-35869-0 (hbk)
ISBN 13: 978-1-138-35871-3 (pbk)
ISBN 13: 978-0-429-43419-8 (ebk)

THE ECONOMICS OF WATER AND WASTE IN THREE AFRICAN CAPITALS

Contents

Figures and tables

Conversions

Conversion to per-capita figures

Whenever per-capita figures are given for a particular year, the city population has been estimated by least-squares regression of the natural logs of the census population (and some intercensal population estimates) on the calendar years.
The coefficients of the regressions are:

	Accra	Harare	Gaborone
Constant	-77.3	-76.2	-221.1
	(0.3)	(0.1)	(0.4)
Year	0.0459	0.0452	0.1170
	(0.0150)	(0.0041)	(0.0144)
R^2	0.82	0.98	0.94

(Standard errors of the coefficients are in parentheses.)

Conversion to US dollars

Throughout, all money values are converted into 1994 US dollars. The country's own Consumer Price Index (CPI) is used to convert the value to 1994 prices and the average 1994 exchange rate with the US$ is then used to convert the value to 1994 US$. The data are given below:

Consumer Price Index (CPI)

	Botswana	Ghana	Zimbabwe
1977	26.5	0.7	20.7
1978	28.9	1.2	21.8
1979	32.2	1.9	25.8
1980	36.7	2.8	27.2
1981	42.6	6.1	30.8
1982	47.5	7.4	34.0
1983	52.4	16.5	41.9
1984	56.9	23.1	50.4
1985	61.5	25.4	54.7
1986	67.7	31.7	62.5
1987	74.3	44.3	70.3
1988	80.5	58.2	75.5
1989	89.8	72.9	85.2
1990 (base)	100.0	100.0	100.0
1991	111.8	118.0	123.3
1992	129.8	129.9	175.2
1993	148.4	162.3	223.6
1994	164.1	197.5	273.3
1994	Pula/US$	Cedi/US$	Z$/US$
Exchange Rate	2.69	1052.6	8.39

Source: IMF, various issues, *International Financial Statistics*.

Whenever the year to which an own-currency figure refers is not reported in a document, we have assumed it to be the year before the publication of that document.

1 Introduction

The three cities

Accra (Ghana), Harare (Zimbabwe), and Gaborone (Botswana) — the three African cities whose urban environment will be closely examined here — are similar in many ways.

First, each is effectively the capital city of a former British colony.[1] What this means in practical terms is that each inherited at independence a strong, if often misdirected or underfinanced, concern for the public health of the cities.

Second, the three cities are alike in that the countries have made a serious and largely successful effort to provide the public goods necessary for the well-being of their urban population. Observation shows this and statistics reinforce it. In Table 1.1, several "human development indicators" are shown for the three countries and for all of Sub-Saharan Africa (SSA).[2] For almost all such indicators, all three countries are above, often far above, the average for all SSA.

But these three countries, and their capital cities, differ in important ways, too. Most noticeable is in the level of GDP per capita. The standard of living in Ghana is about one fifth that in Botswana, and the standard of living in Zimbabwe is about one half that in Botswana (see Table 1.1). The growth rates of GDP per capita have also differed greatly. GDP per capita in Ghana and Zimbabwe has been in almost steady decline since Independence. In Ghana, the standard of living has fallen from being one fifth above South Korea's in 1960 to being about one tenth that of South Korea in 1991.[3] And Zimbabwe's standard of living was the highest of the three countries as recently as the 1970s. Until 1970, Botswana was the poorest of the three — and indeed, one of the poorest in the world — but it has raised its per capita

Table 1.1
Human development indicators

	Botswana	Ghana	Zimbabwe	SSA[1]
Life expectancy[2]	64.9	56.0	53.7	51.3
Adult literacy[3]	67.2	60.7	83.4	54.4
Primary enrollment[4]	119	77	117	66
Years of schooling[5]	2.5	3.5	3.1	1.6
Access to water[6]	100	93	95	73
Access to sanitation[6]	91	64	95	59
Infant mortality[7]	43	81	67	97
Real GDP per capita[8]	4690	930	2160	1232

Notes

1   SSA = Sub-Saharan Africa (population-weighted averages of countries for which data are available).
2   Life expectancy at birth (in years, in 1992).
3   Adult literacy rate (as % of all adults, in 1992).
4   Primary school enrollment (as % of age-cohort, in 1991). Figures greater than 100% indicate that people outside the age-cohort are attending.
5   Years of schooling (average received by persons over age 25, in 1992).
6   Access to water or sanitation (urban, as % of urban population, in 1988-93).
7   Infant mortality rate (per 100,000 live births, in 1992).
8   Real GDP per capita (in purchasing power parity [PPP] US$, in 1991).[4]

Sources: UNDP, 1995, Tables 2, 6, and 16; UNDP, 1994, Tables 2, 10, and 11; World Bank, 1994c, Table 28.

GDP at a rate of 8% per annum over the last two decades, through wise allocation of its large diamond revenues. These different paths of GDP per capita, as we shall see, have brought about tremendous differences in the development of the urban services of these three capitals.

The rates of population growth of the cities themselves have also been quite different. The populations of Accra and Harare have grown slowly — at less than 5% per year over 1960-94 — though for different reasons.[5] Accra's slow growth reflects the slow growth of the economy as a whole and the emigration of urban residents seeking better living standards in other countries; Harare's slow growth reflects the "influx control" policies of the white government before 1980.[6] Gaborone grew at nearly 12% per annum during this period.

The rainfall endowments of the countries also differ greatly, with obvious effects on the urban availability of drinking water and the handling of wastewater. Ghana has much more water than the other two countries — its Lake Volta is one of the largest lakes in the world — but it supplies the least water to the residents of its capital city. Botswana and Zimbabwe are arid countries, and this is reflected in their water and wastewater policies, but these policies are very different between the two cities.[7]

As we examine the urban environments of these three cities (in the next three parts), we shall focus intensively on the public policies that determine who gets what, how they get it, and what price they pay. Before starting that examination, however, it is important to note carefully why freely operating private markets cannot be expected to provide optimal urban environments (or optimal quantities of urban environmental services). This listing of possible market failures will help to organize thoughts about the kinds of public policies that are needed.

Urban amenities, market failure, and public policy

An almost inevitable byproduct of economic development is the movement of people into cities. While the migrants move of their own free will, because they believe they will become better off in the cities to which they move, a second byproduct of development is the increase in the number of poor people in cities. Because the rich will always find adequate environmental services, and the poor usually will not if public policies fail, we will focus on the provision to the poor in our examination of the three cities.

The poor in the cities of developing countries lack many things, starting with food and clothing and going on through housing, health care facilities,

3

and schooling for their children. There is a temptation for a government to try to provide all of these or none of these, depending upon its social philosophy.

Both of these temptations should be resisted, for there are some things that the government can do no better than the market is already doing, and there are some things that the government must do or doom the urban poor to reliance on inadequate and costly private-sector provision. Markets fail for a number of reasons, and many of these causes of market failure attach to the provision of urban services in a way that they do not attach to the provision of food and clothing.

Market failure in the supply of such services as drinking water, sewage treatment, and solid waste disposal can occur for any of the following reasons:

1   *Externalities.* The purchase and consumption of a commodity may impose costs on others (negative externalities) or provide benefits for others (positive externalities). Such external costs or benefits are usually ignored when consumers make their decisions on how much to buy, and hence too much is consumed of products with negative externalities and too little of products with positive externalities. Consumption of adequate water and the purchase of appropriate human and solid waste disposal provide positive health and amenity externalities to the neighboring community and therefore need to be encouraged by public policies.[8]

2   *Public Goods.* There are products for which one person's consumption does not diminish another person's enjoyment (non-rivalness) and for which it is difficult to deprive people who will not pay (non-excludability). While there exist few "pure" public goods, a healthy urban environment comes close to being an example of one. One person's not being exposed to disease and discomfort does not diminish the ability of the next person not to be exposed — indeed it enhances the next person's consumption of "non-exposure." Moreover, it is almost impossible to prevent anyone who does not pay from consuming such non-exposure.

3   *Merit Goods.* There are products that the community feels all its members should consume, at least in some minimum quantity, regardless of their ability to pay or, indeed, regardless of their willingness to pay.[9] Merit goods are sometimes goods with which people are endowed and may be tempted to sell, such as their bodies or their rights, and public policy usually forbids such sales. In the present context, merit goods are goods with which people are not endowed, and public policy forces, encourages,

4

or subsidizes their purchase. Minimal water consumption and waste disposal are often considered merit goods.[10]

4   *Natural Monopolies.* There are goods whose provision requires high average fixed cost and low, or decreasing, average variable cost. When several firms compete in the production of such goods, the high fixed costs are unnecessarily multiplied; but when only one firm produces the good, monopolistic exploitation of consumers will result. The solution to this dilemma is often to provide the good through a single, public (or publicly regulated) firm whose explicit objective is maximum welfare rather than maximum profit. Drinking water and sewerage systems, with treatment plants of large minimum efficient scale (MES) and extensive pipe networks, are examples of natural monopolies.

It is no surprise, therefore, that we find city governments much more involved in the provision of water and the disposal of waste than in, for example, food or clothing. It is in water and waste that private markets are most likely to fail. What is perhaps surprising is that it is not easy, as we shall see, to know what kinds of public policies in water and waste "maximize welfare."

The chief reason why "maximizing welfare" is a complex problem is that the *best* distance between two consumption points is rarely a straight line. The residents of poor cities, such as we will study, currently receive very different kinds as well as different quantities of water, sewerage, and solid waste services. Ultimately, successful development will mean that all urban residents will receive ample quantities of potable, piped, in-house water, will have water-flushed, sewered toilets, and will receive regular curbside pickup of their solid waste. In the short run, however, this is not attainable because the poorer residents are not able (or not willing) to pay for such services and the government is budgetarily not able (or not willing) to provide the subsidies that would be needed. Governments cannot just think of how to expand the fraction of the residents receiving the ultimate first-class service levels; they must also think about what kinds of second-best services are temporarily appropriate for the poor and what kinds of subsidies (if any) are warranted.

In Parts 2 through 4, we shall look closely at each city, in turn, examining who gets what kind of drinking water, wastewater treatment, and solid waste disposal, how much they get, and how the costs of these services are covered. The order of the three cities is arbitrary, from poorest to richest, and the three parts are independent of each other. Finally, in Part 5, we compare the policies, problems, and achievements of the three cities and return to the difficult question of how to "maximize welfare" in the provision of urban environmental services.

Notes

1   The word, effectively, is included since only one of the three countries was technically a British colony. Zimbabwe, as Southern Rhodesia, was a largely independent self-governing territory until 1965 when the white government announced a Unilateral Declaration of Independence (UDI), which lasted (though little recognized by other nations) until a negotiated Independence in 1980. Botswana, through agility in negotiation and luck that the diamonds had not yet been found, avoided colonial status, becoming instead a British Protectorate (Bechuanaland) until 1965.

2   The term "human development indicators" was popularized by the United Nations Development Program (UNDP) in 1990. Its cornerstone — the "human development index" (HDI) — consists of a complex but roughly equal weighting of life expectancy, adult literacy, and GDP per capita (UNDP, 1995, pp. 134f). In 1992, this HDI was .763 for Botswana, .482 for Ghana, .539 for Zimbabwe, and .382 for a population-weighted average of all other SSA countries (ibid., Tables 1 and 16). The gender-related development index (GDI), which includes the same variables but weights them by gender inequality, is even higher for these three countries relative to the rest of the SSA countries: the GDI for Botswana is .696, for Ghana is .460, for Zimbabwe is .512, and for all other SSA countries is .335 (ibid., Tables 3.1 and 16).

3   The 1960 comparison is of GNP per capita in US$ (World Bank, 1980, Table 1). The 1991 comparison is of real GDP (PPP) per capita (UNDP, 1994, Table 4).

4   The World Bank's figure for the real GDP per capita (PPP) of Ghana was increased to US$2110 in 1992 — an increase of 127% in one year. Casual, albeit largely urban, observation strongly urges that the 1991 figure is the more accurate.

5   For SSA as a whole, cities grew at more than 6% per annum in this period (Becker et al., 1994, pp. 32f).

6   "Influx control" was apartheid's attempt to prevent male black workers from entering the cities until they were needed in the workforce (and to prevent females from entering at all).

7   One indication of the different policies is the fact that the average domestic consumption of water in Zimbabwe, 52 liters per capita per day,

is nearly four times that of Botswana, 14 liters per capita per day (World Bank, 1994c, Table 33). (Five liters per capita per day is considered the biological minimum needed to sustain human life.)

8   The externalities are not all positive. Greater water consumption, for example, means greater wastewater generation. Since the water consumer does not usually pay the full cost of the disposal of this wastewater, there is an element of negative externality in greater water consumption.

9   There are thus two aspects to the "merit good" argument for public provision: 1) that consumers ought to get more of what they want but cannot afford, or 2) that consumers ought to consume more of what they can afford but do not want (presumably because they fail to recognize the value of the consumption to their long-term well-being). The first seems more appropriate in the present context.

10   It is also possible to view merit goods as a subset of public goods. For example, "knowledge that everyone in the community is receiving a minimum or an adequate supply of something generally considered essential for well-being" can be thought of as a "good" that is non-rival and non-excludable.

2 Accra

with Louis Boakye-Yiadom Jr.

Introduction

Accra is a vibrant city in the classical mold. At its core, there are modern commercial buildings, traditional government structures, and old colonial-era residences, now densely packed with the poor. And sprawling out from this central business district (CBD) are the residential suburbs, regularly interspersed with open spaces and usually segregated by income class (Stephens et al., 1994, p. 20; World Bank, 1994a, p. 24).

Accra is not a typical African city. A quarter century ago, Ghana was the most urbanized of all West African countries, and its urban population growth rate was the highest, some 10% per annum (Gugler and Flanagan, 1978, p. 38; World Bank, 1989a, p. 4). After Independence in 1957, however, with almost continual economic mismanagement and stagnation, the urban population growth rate fell to barely 3%, the lowest in Africa.

Thus, Accra was spared the infrastructure problems that accompany extremely rapid urban population growth. But Accra did suffer under two other problems.

The first problem arose from the economic decline. While it slowed up the growth of Accra's population, it even more seriously damaged Accra's ability to afford the growth that did occur. Tax revenue suffered at all levels of government, compromising first the funds for urban infrastructure investment and then even the funds for infrastructure maintenance and repair.[1] Also, the economic decline has caused many of Ghana's high-level personnel to emigrate, leaving both the Accra Metropolitan Authority (AMA) and the Ghana Water and Sewerage Corporation (GWSC) with grossly inadequate technical and managerial competence (GOG, 1993, p. 33; World Bank, 1993a, pp. 73f). As a result, almost no recent discussion of Accra's infrastructure

8

seems to avoid the word "deterioration" or "disgrace" (World Bank, 1989a, p. 23; GOG, 1992a, p. 1; World Bank, 1993a, p. 73; GOG, 1993, pp. 5, 27; GOG, 1995, p. 15). Now, it seems that no more than "rehabilitation" is sought by policy proposals:

> The overall thrust of the services strategy in the short and medium term should be directed to the continued rehabilitation and improved utilization of engineering services .... seek to establish a basic level of service .... These may be below public expectations, but services will be affordable, functional, and maintained. (GOG, 1992c, p. 82)

The second problem with the provision of urban infrastructure arose from the physical dimensions of Accra's growth. Ordinances designed to give sensible structure to Accra have "for several decades not been enforced ... [and] spontaneous developments have characterized the sprawl in many parts of the city" (Akuffo, 1989, p. 2).[2] Not only has Accra sprawled, but there has simultaneously been "increased crowding in existing residential areas ... particularly severe ... [in] unserviced and unplanned slum areas" (Benneh et al., 1993, p. 7). Today, two thirds of Accra's households live in "single story traditional house compounds, occupied by several households, and often sharing sanitary and kitchen facilities," with 4.2 households, or 23.3 persons per dwelling (ibid., p. 9). In these low-income areas of Accra, density often surpasses 300 people per hectare and the infrastructure is "inadequate," "deteriorated," "overstretched," and/or "lacking" (GOG, 1992a, pp. 1, 10; GOG, 1992d, p. 82).

Faced with limited budgets and demands for extensive servicing by the rich and intensive servicing by the poor, the AMA and the GWSC have generally handled the rich less badly than the poor (GOG, 1993, p. 30).

Drinking water

Accra draws its water supplies mostly from Lake Volta, which provides an ample, cheap, and unpolluted water source for the city and its environs.[3] But the drinking water that reaches most of Accra's citizens is not ample, cheap, or unpolluted.

The drinking water problems of Accra are not brought about by the low priority that the government places on this service. From its independence, the provision of drinking water has been a high priority. The first post-Independence plan document stressed that the Government attaches great importance to water supply and sewerage. ... The ultimate aim of the domestic water supply service in Ghana is to provide good and abundant water to all parts of the country and to achieve house-to-house delivery, thereby eliminating the public stand-pipe system of supply, since this detracts so much from the value of water service by the element of pollution it introduces. ... The Kpong extension and Akosombo development ... will increase the water supply in Accra and Tema ... to 40 mgd [million (British) gallons per day], the amount judged to be needed in this area for all purposes including sewerage. (ROG, 1964, pp. 131f)

And subsequent plans reaffirmed this emphasis:

... the Government will give first priority to the expansion [of clean water supplies]. (ROG, 1968, p. 92)

And

... existing urban water supply schemes will be expanded to meet the rapid increase in economic activity .... (ROG, 1977, p. 483)

In actuality, recorded water sales in Accra by 1993 were still less than 30 mgd, while the city's population had increased fourfold in the intervening 30 years.[4]

Most Ghanaians draw their water from rivers and wells, but in Accra the population relies almost entirely on the public piped water supply (Huq, 1989, pp. 59f; Benneh et al., 1993, p. 13). What differs among residents is how much they get and how they get it.

More than half of Accra's residents get their water either through indoor plumbing or through a private in-yard tap. Exactly how much more than half is unclear.[5] And it may well be that the fraction has declined somewhat over the last decade (World Bank, 1994a, p. 11). The rest of the City's residents get their water from a variety of sources: water vendors, communal standpipes, and neighbors. Again, the fraction who utilize each of these sources is unclear. Some studies claim that neighbors' sales are the primary source of water for those without their own taps (GOG, 1992b, p. 29); other studies claim water vendors (Ankrah, 1994, pp. 16f; Benneh et al., 1993, p. 11; Songsore, 1992, p. 5); what is clear is that communal standpipes serve no more than 10% of the population — and perhaps as few as 1% (GOG, 1992a, p. 104).

10

How one gets water is closely related to how much water one gets; there are two reasons for this relationship, one tied to price elasticity and the other tied to income elasticity. The price of water is generally lower for those who have in-house water taps; and higher-income households are more likely to have in-house water. Both of these phenomena deserve closer examination.

Pipes are the cheap way to transport water. When the poor buy from vendors — who move water by hand-cart or tanker-truck — they pay much more than they would if they were to purchase piped water from the GWSC.[6] For those who buy from neighbors or communal standpipes, the water may (or may not) be reasonable in money cost but it involves a time cost as well.[7]

Given that the more convenient water source, an in-house tap, is cheaper, one would expect even poor households to prefer it. Unfortunately, there is a sizeable cost attached to the initial tap connection. The exact cost depends upon the distance to the water main, but it typically runs around $100, no trivial amount for a poor family in Ghana.[8] The result is that almost all high-income households have in-house water taps, and some 85% of the middle-income households do, while barely half of the low-income households have in-house or in-yard water (Benneh et al., 1993, p. 12).

In short, the poor in Accra get much less water and pay a much higher percentage of their incomes for it than middle-income and high-income households. Studies vary, but the consensus is that high-income and middle-income families with in-house connections consume 100-150 liters per capita per day (l/c/d) of water, while low-income households consume 30-60 l/c/d (Akuffo, 1989, p. 7; Songsore, 1992, p. 6; Tahal, 1980, p. III-7; GOG, 1988, pp. 62ff). While the well-to-do rarely spend more than one or two percent of their incomes on water, the poor spend as much as 10-24% of their incomes (Benneh et al., 1993, p. 19; Tahal, 1980, p. III-6).

Moreover, the quality of the water received by rich and poor in Accra is not the same, for two reasons. One, while the treated water from Lake Volta (and other sources) is quite potable, once that water has been carried and stored by those without in-house supply, it is no longer unpolluted water. And two, the water from the taps in poor areas of Accra is often locally contaminated because of "illegal and improperly fitted connections, breakages and loose joints of pipes which run through cesspools and gutters" (Benneh et al., 1993, p. 23).

Given that pipes move water more cheaply and more cleanly than trucks, carts, or legs, the poor in Accra would get more, cheaper, and perhaps better drinking water from a widespread and functioning system of communal standpipes. The number of public standpipes provided by the GWSC has, indeed, increased dramatically over the past few years, from 100 in 1988 to

348 in 1994, but the number had fallen almost as dramatically in the early 1980s (from 275 in 1980, according to Tahal, 1980, p. III-5).

The GWSC, however, has never considered public standpipes as a permanent and efficient part of its water delivery system, a means of providing ample quantities of drinking water to the poor at prices they can afford. Traditionally, whenever a new residential area has appeared, the GWSC has installed standpipes to service the new residents. These standpipes were not superintended and not metered, and hence they tended to break down, waste water, discourage in-house connections (which were metered), and cost GWSC money. As a result, not surprisingly, the GWSC could not wait to reduce the number of such standpipes over time as the residents became connected up to in-house water systems. Indeed, the GWSC may even have let the standpipes disappear a little too rapidly, by failing to repair economically efficient but out-of-order standpipes, in order to encourage such in-house connections.[9] Thus, those who chose not to connect, or could not afford to connect, had to make do with ever fewer and ever more distant standpipes.

More recently, the GWSC has begun to meter and supervise its standpipes in new residential areas (GOG, 1992d, p. 122).[10] But it has not aggressively recruited operators for existing standpipes, and it continues to believe that developers should install in-house connections for all. What the GWSC must realize is that, for those too poor to find (first-best) in-house connections desirable or feasible, the absence of (second-best) standpipes forces them to (third-best) vendor or neighbor purchases.

All governments hope that all their citizens will some day have in-house water supply. But this hope must not stop interim efforts to provide water by less luxurious means. Having a nearby standpipe does not assure high-quality water, but it can assure plentiful and low-cost water, and much current thinking stresses quantity above quality of water for maximal health impacts. For a long time, planners in developing country cities have recognized that standpipes may be *necessary* as an interim measure.[11] What is too often not recognized is that standpipes may be *desirable* as an interim measure.

Price, quality, proximity, and convenience aside, the question remains: if there is ample water in Lake Volta, why is there not ample water in Accra? Part of the answer is that we do not know how much water there is in Accra — much of what is produced (i.e. treated) is simply unaccounted for, in the sense that its precise destination is unknown. Estimates of the extent of this unaccounted-for water (UFW) vary a great deal, but all the estimates are large.[12] Water "disappears" between being treated and being paid for in many ways: 1) the meters at the treatment plants are faulty and may overcount the amount of water actually treated;[13] 2) transmission lines to Accra leak;[14] 3) water may be illegally removed from the main transmission lines for agricultural purposes; 4) pipe leakage occurs within Accra; 5) there are illegal connections in Accra;[12] 6) faulty (or tampered with) meters undercount water consumption;[13] 7) many consumers are unmetered;[14] and 8) consumers do not always pay their water bills.[15]

UFW is always a social problem in the sense that it means costs are being incurred while revenues are not, exacerbating the seemingly ubiquitous budgetary problems of the water authority. But of the four major sources of UFW in Accra — transmission losses, unmetered customers, illegal connections, and unpaid bills — only the first is a total social loss; the others represent unplanned income redistribution. This redistribution means that at least some of Accra's poor are getting more water than the statistics show, but the budgetary impact of UFW means that the GWSC is even less able to undertake planned redistributive water policy.

In addition to losses on the way, water flows to Accra at intermittent rates and at variable pressures. The problem begins at the treatment plants, where the "obsolete equipment" means that supplies are "often disrupted" (Benneh et al., 1993, p. 14). Within Accra, "inadequate main secondary and tertiary pipelines" mean low pressure and interrupted supply for many areas (GOG, 1992d, p. 120). Some new areas of Accra receive inadequate water "because the rate of housing development far outstrips the mains extension capacity" (ibid.). In low-lying residential areas, any reduction of pressure may let polluted groundwater enter the system. On the uphill (west and north) sides of Accra, frequent breakdowns of the old and unmaintained pumps leave huge areas without water for extended periods — Madina, a particularly poor section of the City, is "constantly without water" (Ankrah, 1994, p. 14). Benneh et al. (1993, p. 14), estimate that two thirds of Accra's households suffer "a regular daily interruption of their principal drinking water supply."[16]

The social costs of intermittent supply and erratic pressure cannot be overstated. It forces people to store water.[17] This means container costs, time costs, inconvenience costs, and costs in foregone water quantity and quality. It also forces the City to maintain a fleet of water tanker-trucks to assure that minimal supplies of water reach all citizens.[18] This, too, is costly since trucks are an inefficient way to move water, especially where pipes already exist. Moreover, for many households, it means long walks for water — especially long walks given the GWSC's policy of gradually removing standpipes in areas that are, in principle, receiving piped water. Finally, it should be noted that all these problems stemming from low pressure and erratic supply have greater impact on the poor than the well-off; the well-off usually have more

13

and cleaner in-house storage facilities and better access to delivery of clean water by tanker-trucks.[22]

This is the real tragedy of Accra's water system. The total social costs of the existing intermittent, low-pressure system *vastly exceed* the total social costs of operating an efficient system. New pumps, regular pipe inspection, and regular maintenance throughout the system would provide not only better water delivery but also cheaper water delivery.

The reason for the existing water system problems is not hard to find: while the social costs of operating it are high, the actual money costs are low. Indeed, as Figure 2.1 shows, the money cost of the GWSC Accra operation runs around $4 per capita per annum (in 1994 U.S. dollars). Furthermore. as Figure 2.1 also shows, the excess of income over operating expenditure never exceeded $5 per capita per annum until 1993, leaving little surplus for investment.[23] Finally, with new and ever more distant suburbs demanding water service each year, there is even less left over for discretionary investment.

Many have stressed the need to cut unnecessary cost in the GWSC, and to improve the managerial and engineering services.[24] What is also needed are water tariffs that not only cover recurrent costs but also contribute to the maintenance and expansion of the system. Real water charges have been generally declining in Accra for the past decade, as Figure 2.2 shows.[25] For families consuming more than ten cubic meters of water per month, the real payment has rarely been as high over the past ten years as it was in April 1984. While the real cost of water for a low-income family of seven members, consuming just under 50 liters per capita per day (l/c/d) of in-house piped water, rose dramatically in March 1986, it also has fallen by more than 40% since then — from $1.50 in March 1986 to $0.87 in December 1994. GWSC has not pursued tariff increases aggressively enough. And the government has worsened the situation — sometimes it has "actually disallowed increases in tariffs" (Engmann, 1990, Water Supply, p. 3) — even though it has long preached that water should be paid for at a rate that covers costs (ROG, 1959, p. 52).

Human waste

Human waste falls into two categories, sullage and septage. Sullage, the waste water of washing, is in Accra almost entirely discharged directly into yards, drains, ponds, and rivers.[26] Septage, the waste of toilets, is handled in a variety of ways in Accra (GOG, 1988, pp. 72ff). These septage collection processes

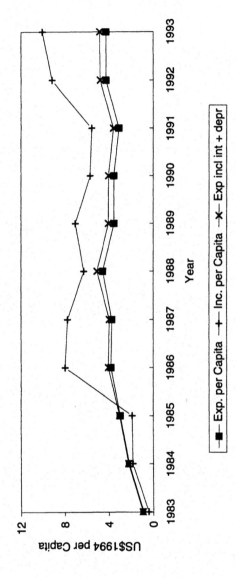

Figure 2.1. Water income, expenditure (in US$ 1994 per capita)

15

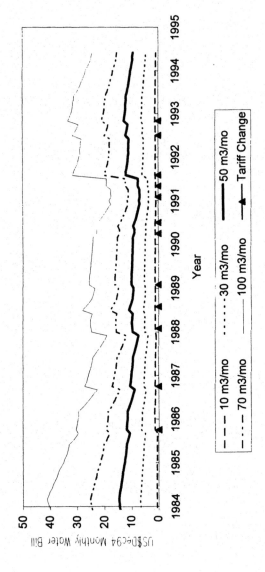

Figure 2.2. Monthly water bills, 1984-95 (in US$ Dec 94 per month)

16

are more sanitary and less unsightly than those of sullage — as they should be, given the additional health risks that septage poses — but, once collected, much of the septage is then discharged without any treatment.

Sewers are very efficient ways of handling septage, but they are expensive, require that all households have ample water, and concentrate the effluent in such quantities as to demand treatment. Accordingly, in Accra, many other disposal processes are also used. Table 2.1 shows the principal means, for both the beginning and the end of the 1980s.[27] The two sets of surveys were not comparably collected, and may well yield incomparable figures, but they do suggest two things: 1) little changed in the 1980s in the distribution of the types of toilets used; and 2) there was a significant increase in the usage of public facilities.

As one might guess, the type of toilet used in Accra differs by income group (Table 2.2). High-income households usually have in-house water distribution, usually have flush toilets, and usually discharge the effluent into a septic tank. Middle-income and low-income households are more likely to rely on pit latrines or pan latrines — or "none." In the remainder of this section, we will look, in turn, at the processes and problems of each of these disposal techniques — sewers, septic tanks, pit latrines, and pan latrines — as well as at public toilets.

Even before independence, the need for a comprehensive sewer system for Accra had "long been recognized", and the early five-year plans made it a high priority, promising "strenuous efforts ... every encouragement ... [and] foreign exchange cover" (ROG, 1959, p. 53; ROG, 1968, p. 93). With World Bank assistance, the GWSC undertook this "first phase" of Accra's water-borne sewer system; it was completed in 1973, covering 1,000 ha and involving 28.5 km of sewers (GOG, 1992d, p. 127; GOG, 1988, p. 17).

Unfortunately, this proved to be a classic example of aiming for the first-best solution before either the government or the potential beneficiaries could afford it. The system was very expensive, it never benefited very many people, and it never worked very well.

The sewerage system was installed in central Accra, where the housing is dense, the population is crowded, and the streets narrow and crooked. One observer noted that the "houses and plumbing in most parts of the area will never meet the standard for connection to the system" (Akuffo, 1989, p. 45), which means that the sewer should have been postponed until a radical redevelopment of the City center was feasible.

Beside the inability of some houses to connect, houses that could have been connected to the system were choosing not to be connected. For the rich, their houses already had adequate septic tank systems; and for the poor, the

17

Table 2.1
Distribution of Accra households by type of toilet used

Type of Toilet	Year 1980	Year 1989
Flush Toilet	30%	31%
Pit or Pan Latrine	62%	56%
None	8%	9-13%
Public Toilet	18%	25-59%

Sources: For 1980 — Akuffo, 1989, p. 6; Tahal, 1980, p. IV-9. For 1989 — ROGSS, 1995; GOG, 1992c, p. 89; GOG, 1992a, pp. 110ff; GOG, 1994, p. 36; Stephens et al., 1994, p. 25; World Bank, 1994a, p. 11.

Table 2.2
Distribution of Accra households by wealth and type of toilet used, 1991

Type of Toilet	Income Low	Income Medium	Income High
Flush Toilet	26%	68%	98%
Pit Latrine	48%	16%	2%
Pan Latrine	22%	11%	0%
None	4%	5%	0%

Source: Benneh et al., 1993, p. 28.

connection fee was not trivial, and those who became connected then started receiving a 25-35% surcharge on the water bill.[28] As of 1980, there were "only some 220 connections," and those mostly commercial premises (Tahal, 1980, p. 25). By 1991, still only 725 households had been connected (GOG, 1992d, p. 127).[29] The number of connected households has doubled in recent years, partly because the City has announced a ban on the construction of pit latrines in central Accra (Stephens et al., 1994, p. 25).

Operating below designed capacity has created problems of insufficient flow, which means that the sewer is not self-cleaning (GOG, 1992d, p. 127). This creates a need for costly jet-flushing. Add to this that the jet-flusher is "currently out of service", that "the pumps have broken down", that there is no treatment of the effluent, that the submarine outfall pipe, supposedly carrying the effluent out more than a kilometer into the ocean, is "rusted, blocked and abandoned", and hence that the raw sewage is discharged "just at the beach" (GOG, 1992d, pp. 127f, 132; GOG, 1988, p. 19; Akuffo, 1989, p. 21).[30]

What can we learn from this disaster? Two things. One, the problem is not, as many seem to think, that the system gets "poor cost recovery" (Amuzu and Leitmann, 1991, p. 27) — the revenues do in fact cover the inefficiently inadequate operating expenditures.[31] The problem is that sewer systems can rarely recover their full operating and capital costs if connection is voluntary. Even for sewer systems where the benefits exceed the costs — which almost certainly was not the case here — the benefits involve so many external benefits that recipients must be forced onto the system or big subsidies must be provided. In the short run, Accra might belatedly learn something of this lesson by charging the connection fee and the sewer surcharge to all houses in the area served by the system, if connection is feasible, *whether or not* they connect.

The second lesson is that sewers are very expensive, and that cheaper means of dealing with septage are necessary until Ghana reaches a higher standard of living. In many parts of Accra, sewers are not feasible until houses and streets are redesigned, and the residents are not yet rich enough to make that a priority expenditure; and in parts of Accra where sewers are now feasible, installation might nevertheless have to take place at a loss, which would mean subsidies to the relatively better-off.

For houses with water but not attached to the sewer system, the waste goes to a septic tank; and for houses without water, some kind of pit or pan latrine is used. None of these is ideal in an urban setting — they are not as effective in reducing disease, they are aesthetically displeasing, and they endanger the groundwater. But all may play a useful interim role as a city grows and its citizens grow wealthy enough to afford full sewers.

All of these alternatives to sewers as means of disposing of human waste require periodic cleaning — the septic tank must be suctioned about once every two or three years, the pit must be emptied about once or twice a year, and the pan requires nightsoil collection two or three times a week. These frequencies apply with normal use. But toilets in Accra do not get "normal" use. Families — indeed, entire blocks of flats — often share one toilet.[32] In the "indigenous areas" of Accra, there are 30-35 persons per toilet (GOG, 1992d, p. 90; GOG, 1992c, p. 88). Such heavy use does not mean more total effluent, but it reduces the incentive for people to care for their facilities, and it increases the urgency of timely emptying services.

It is conventional to complain about these cleaning services — e.g. "the current operation of septic tank collection services in Accra ... is very inefficient" (GOG, 1992c, p. 94) or "service for emptying these [pan latrines] has been a bugbear" (GOG, 1988, p. 33). In fact, however, they seem to be reasonably efficient. Figure 2.3 shows the volumes of human waste collected by Accra's Waste Management Department (WMD) and, for the last three years, by private haulers of septic tank sludge.[33] While the trends seem, unfortunately, downward, something like 90% of the human waste generated in 1992 was collected.

Septic tank cleaning is done by three different groups: 1) the WMD; 2) large government and private organizations that buy their own suction trucks for servicing their offices, factories, and workers' housing; and 3) several private firms.[34] The WMD and the private firms charge the same fee.[35] All the septic tank sludge is then trucked to one of the two City sewage treatment areas, where the government and private trucks pay a tipping fee of about $5 per truckload, and the sludge is made into fertilizer, which is sold by the WMD for about $60 per "truckload."[36] Between 1) the charge for septic tank cleaning, 2) the tipping fee at the treatment plants, and 3) the price of the fertilizer, the WMD's septic tank operation is supposed to break even — indeed, the City's fee is supposedly set in order to achieve a break-even position (Pokoo, 1994, p. 16; World Bank, 1994a, p. 32; GOG, 1992d, p. 128; Benneh et al., 1993, p. 30; Darkwa, 1990, p. 20).

Casual inspection of the WMD accounts for one month, however, suggests that the operation does not break even. In January 1995, billings for septic tank cleaning were only $9 thousand, while operating costs were $19 thousand (WMD, 1995a, pp. 3f).[37]

It is not true that "private haulers have not been able to compete with government owned and operated trucks due to the subsidization of public sector equipment and the high tipping fee (relative to the standard emptying charge) charged to private operators" (World Bank, 1994a, p. 32). While the

20

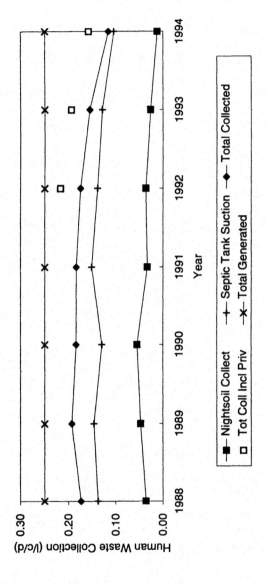

Figure 2.3. Human waste collection (liters per capita per day)

21

price charged by the WMD is probably subsidized in that it does not cover depreciation or interest, the inadequate numbers of City suction trucks — never more than twelve over the past decade and usually several out of action — has meant long delays in WMD service and under-the-counter payments, all of which has left ample scope for private operation.

For families using latrines, the WMD encourages the construction of Kumasi Ventilated Improved Pit (KVIP) latrines, which better control odor and flies, are easier to desludge, and require less frequent cleaning. This encouragement takes two forms: 1) the construction of new pan latrines has been banned "anywhere in the city," and 2) the cost of conversion to KVIPs is subsidized (GOG, 1988, pp. 33ff).[38]

Finally, there are some 10-30 thousand pan latrines in Accra.[39] The number of pan latrines is probably decreasing, not only because the City encourages KVIP latrines, but also because 1) landlords are converting bathrooms into rented rooms (GOG, 1992a, p. 38; Pokoo, 1994, p. 37), and 2) new houses are being built without any toilet facility, despite building regulations requiring them (GOG, 1992a, p. 194).

Until recently, pan latrines were emptied by the WMD. "Conservancy laborers" collected nightsoil from households two or three times a week for a price of about $2-3 per month (GOG, 1992d, p. 128; Darkwa, 1990, pp. 22f; Songsore, 1992, p. 6; WMD, 1994a, p. 14). But the WMD, in order to encourage the switch to KVIPs and to escape problems with non-payment, discontinued nightsoil collection at the end of 1987 (Pokoo, 1994, p. 18; GOG, 1988, p. 33). The service is now provided by the private sector — service is good in some areas, "irregular" in others (GOG, 1992c, p. 89).[40]

The conservancy laborers carry the collected septage — often on their heads and often for long distances — to one of the some 50 privately operated nightsoil depots, from which the WMD removes it at a price of about $15 per (eight cubic meter) container (Darkwa, 1990, pp. 42ff; Pokoo, 1994, p. 16).[41] This collected nightsoil is then dumped on the beach near Korle Lagoon, from whence it is (partially) removed by the next high tide (Amuzu and Leitmann, 1991, p. 27; Songsore, 1992, p. 7).

The last means of disposing of human wastes to be examined is the public toilet. In Accra, this has been viewed as last and least. The "first-best" mindset is nowhere more clearly displayed than in Accra: the City would prefer water-flushed toilets flowing into sewers; if that is not possible, private septic tanks or pit latrines are acceptable; then private pan latrines; and distantly last, the public toilet. This preference ordering is quite explicit:

There are no toilet facilities in most houses and public toilets are not well kept and not in general use. The restricted use of public toilets is ... a deliberate AMA policy.... The objective of the AMA is to encourage the construction of private toilets and to stop the construction of public facilities. (GOG, 1992a, pp. 36, 194f)

Nevertheless, for many of the residents of Accra, the public toilet is the preferred means of human waste disposal, given the prices they face and the sizes of their houses and plots.

There are about 150 public toilets in Accra, mostly of the pit latrine type; the number has changed little in the last two decades.[42] This is the number of blocks, and the number of toilet seats per block, run anywhere from 12 to 40 (GOG, 1992a, p. 100). Conservative use of the above numbers tells us that the number of people using each seat averages at least 75, and it may run into the thousands in the poorest, most congested, and most publicly neglected residential areas.[43]

There are too few public toilet facilities in Accra because, in the minds of the City officials, they are supposed to be merely for a transient trade, with each person having his own private toilet for basic use. Their heavy overuse is one reason why they are often unclean, overfilled, or odoriferous. There is also the free-rider problem, that anyone who takes care in using the toilet pays all of that caring cost but gets back only a small fraction of the benefit.

By the mid-1980s, the WMD despaired of keeping up with its public toilets. The operation was turned over to the local area council, the Sub-Metropolitan District Authority (SMDA) (GOG, 1988, p. 32; World Bank, 1994a, p. 34). These local councils knew the users and the users knew the councillors, and the toilets quickly became cleaner. A charge was instituted, currently about $0.02 per use, to cover the Council's added cleaning expenses (Benneh et al., 1993, p. 30).[44]

The WMD performs two services on these public toilets on request from the local council: the WMD desludges the toilets, and it will build more — both at a price designed to cover cost (GOG, 1988, p. 36; Pokoo, 1994, p. 16). The system works much better than it did before. One might say it works well, except for three problems: 1) the WMD has trouble collecting payment for its services from the local councils (Pokoo, 1994, p. 18); 2) the toilets are still too few and emptied too infrequently (GOG, 1992d, p. 130; World Bank, 1994a, p. 32); and 3) the fee for use has deterred some potential users, so that there is still extensive hole-digging, stream-using, and beach-defecating (GOG, 1992a, pp. 33, 195; GOG, 1992a, p. 194).

For the near future, Accra's sewage problems will be best attacked in three ways: 1) forcing attachment to the sewer system of houses that can be readily attached; 2) permitting septic tanks or pit latrines for those people who want them and in those locations where their use will not be environmentally damaging; and 3) making clean, comfortable, convenient, and most of all cheap public toilets available for the rest.

Solid waste

One of the few advantages of being a poor country is that little solid waste is generated. One of the many disadvantages of being poor is that little of that solid waste is properly disposed of. Accra is no exception on either count.[45]

There have been many estimates made of the volume of solid waste produced in Accra, with remarkable consensus. Solid waste runs around 0.5 kilograms per capita per day (kg/c/d) in the residential areas, which waste amounts to about two thirds of the total for the entire city (Oddoye, 1985, p. 35; GOG, 1988, pp. 67ff; Amuzu and Leitmann, 1991, p. 29; Songsore, 1992, p. 13; GOG, 1992d, p. 144; WMD, 1993, p. 9). There is some, though small, income elasticity to this waste generation, with the low-income areas producing only about 0.4 kg/c/d and the high-income areas about 0.6 kg/c/d.

All of the residential solid waste is highly organic, with the putrescible component representing 70-90% of the total (Darkwa, 1990, p. 19; Amuzu and Leitmann, 1991, p. 29; Botchie, 1994, p. 7). This means that it is very dense — something like 500 kilograms per cubic meter (kg/m^3) — which bodes ill for compactor trucks and recycling but well for composting. The solid waste is especially dense in the poorest areas, where it consists heavily of ashes, yard sweepings, and vegetable and fruit peelings.

While these estimates of the total solid waste generated in Accra are modest — about 0.4-0.5 m^3 per capita per year (m^3/c/yr) — *less than half* of it is actually collected and disposed of in landfills. The official WMD figures, shown in Figure 2.4, show the total collection rate to have fallen fairly steadily over the past decade from 60-75% to 40-50%.[46] Other observers agree with these percentages and this downward trend.[47]

Before trying to understand why the system is so incapable of collecting Accra's solid waste, we must look at the collection process itself. There are basically three methods of household waste collection: 1) WMD curbside collection by truck directly outside of each house, for the wealthiest 5-13% of Accra's citizens; 2) WMD collection from communal containers to which people must carry their own waste, for low-income areas; and 3) door-to-door

24

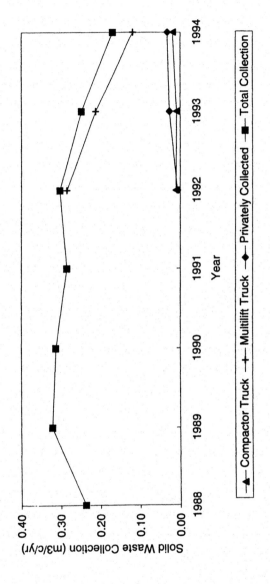

Figure 2.4. Solid waste collection (cubic meters per capita per year)

25

collection in middle-income areas by labor-intensive methods by private haulers under contract to the WMD (GOG, 1992d, p. 144; Benneh et al., 1993, p. 41; Stephens et al., 1994, p. 25; WMD, 1994b, p. 14).

The first collection method, curbside pickup, is provided weekly in the high-income neighborhoods, mostly by the City's three or four operating compactor trucks (Pokoo, 1994, p. 20). The trucks collect from standard-sized containers that are sold to households by the WMD at cost.[48] Residents report that the service is quite timely. For this service, the households are charged; for a house that uses a single 120-liter container, the charge is now $4.40 per month — a sizeable increase from the fee of $1.60 four years ago (Benneh et al., 1993, p. 42).[49]

The bills for this curbside service are sent out quarterly, and collection efforts are made by WMD Revenue Assistants who can earn bonuses for exceeding a target of 80% collection (WMD, 1994b, p. 13; WMD, 1994a, p. 5). Currently, however, the average collection rate is only 40-70% (WMD, 1994b, p. 18; WMD, 1995a, Summary, p. 2).

Even lower than the collection rate is the participation rate: where curbside service is provided, only about 60% of the houses choose to subscribe to it (ibid., p. 19).[50] With fewer than half the houses that the trucks and crews drive by actually paying for the service, it is surprising that it manages to cover operating costs; but this service for high-income neighborhoods contributes little to overheads or other WMD costs (WMD, 1994a, p. 6).[51]

The second collection method, used in the low-income parts of the City, is through some 200 communal containers, to which households may take their refuse and from which the WMD collects it (Stephens et al., 1994, p. 25; WMD, 1994a, p. 4). In principle, the City's 10-15 operating multi-lift trucks visit these seven-cubic-meter-containers, or "skips", as often as needed, but the WMD has not enough trucks for this:[52]

> The containers are supposed to be removed daily, but the usual problems plague these depressed areas and the containers are not emptied promptly and garbage spills over. (GOG, 1992a, p. 36)

This seemingly inequitable difference in collection technique has been justified by the argument that in "the older parts of the city ... households are inaccessible to collection vehicles" (GOG, 1988, p. 32). But it is also defensible on the grounds that the poor may be neither able nor willing to pay for curbside pickup given the price the City charges for that service. The WMD fee for community container service has always been much lower.

In fact, over the past seven years, Accra has tried three different methods of charging low-income households for their use of communal containers:[53]

1 The "refuse levy" (1988-93). This was a tax of $2 per household per year surcharged onto the property tax; it had been increased to $3 by mid-1993, at which time the levy was abandoned (WMD, 1994b, p. 19). Too few bothered to pay. Property taxes are hard to collect anyway in Accra, and some people were even paying the property tax after tearing off the surcharge part of the bill. Only 10% of the relevant households paid in full, and only 30% ever paid anything (ibid.; GOG, 1992d, p. 145; Benneh et al., 1993, p. 42). The levy was hardly enforceable, since the communal containers were there for every household whether it paid or not.

2 "Pay-As-You-Dump" (1993-94). This was a charge of $0.02 for each household container dumped into the communal container ($0.04 if it were very large). These fees accrued to the Sub-Metropolitan District Authority (SMDA), which in turn paid the wage of the operator of the communal container and a fee to the WMD for emptying of the full containers. The SMDAs became derelict in their payments to the WMD; there was a great increase in illegal dumping, and the system never covered even one third of the WMD operating costs.[54] Pay-as-you-dump (PAYD) was dumped late in 1994.

3 SMDA Responsibility (1994-). Each SMDA is now responsible for paying the WMD for each container emptied and for collecting — somehow — the needed money from its residents. Since the abolition of PAYD, the WMD is having more trouble collecting from the SMDAs, and hence more trouble operating its vehicles at the very time that the rate of trash reaching the communal containers is increasing:

Skips placed at various collection centres in especially densely populated areas ... are spilling over with refuse. (Aboagye, 1995 p. 1)

What do we learn from these three failed attempts? Certainly not the lesson the World Bank drew:

The experiments in Accra suggest that the share of the costs recovered from users ... can be raised significantly. (World Bank, 1994a, p. 38)

Rather, the lesson is that it is very difficult — and perhaps impossible in the poor residential parts of a developing country — to charge and collect a price for solid waste collection. It is not that people do not value a neat and sanitary neighborhood. But what they value is the neat and sanitary neighborhood, not the removal of the family's own solid waste. This is a classic public-good problem with classic free-rider incentives.

With public toilets, a small price per use is feasible, since part of what is being purchased is a private good, modesty and privacy in one's own ablutions. But with solid waste, there is almost no element of a private good there, and hence charging a marginal price in poor neighborhoods will always lead to evasion through illegal dumping or non-payment. A periodic lump-sum charge per household is feasible, at least just as feasible as any property taxes ("rates") are, but not marginal charges.

Does this mean that solid waste collection in low-income areas must necessarily be financed from the general fund? To some extent yes, but not entirely. It is always possible to cross-subsidize (partially at least) this activity from surpluses earned in high-income neighborhoods, and these surpluses could be much increased in Accra by requiring participation where curbside service is offered and by strengthening the collection procedures. This is not just an income-redistribution policy — the rich have a willingness-to-pay for cleaner neighborhoods throughout the City.[55]

For low-income neighborhoods to become cleaner requires, as a necessary condition, the abandonment of marginal charges for solid waste in these areas. But more is required. Communal containers must be emptied regularly, and there must be enough communal containers.

Finally, there is private collection of solid waste in Accra. In an effort to reduce the number and area of its communal container pickups, on which the WMD lost money even under the refuse levy and PAYD, the City began in early 1992 to contract out mixed-income and middle-income residential areas to private waste collectors (WMD, 1994b, pp. 34f; WMD, 1995b). The City and the contractor negotiate the area, the service to be provided, and the rates to be charged. The City monitors the operation and charges tipping fees at the landfill, but it is up to the private contractor to collect the household fees in its service area. The contractors typically provide a mix of curbside pickup and communal container service, letting each household choose which service it wants. By mid-1994, there were nine of these private contractors, covering about one fourth of the City's houses and collecting about one fourth of the City's solid waste (WMD, 1994b, pp. 22, 30; WMD, 1995a, Solid Waste Collection Report 1).

The private contractors typically charge a fairly low basic rate for curbside pickup — currently around $2.50 per month, compared to the WMD's $4.40 per month (WMD, 1994b, p. 23).[56] The private contractors are able to thrive at such low charges because they get high participation ratios and high collection ratios and because they operate cheap, labor-intensive equipment — pushcarts, donkey carts, open-body trucks, tractor-trailers, and old tipper trucks (ibid., pp. 58f).[57]

The success of the private sector here points up the failure of the public sector. Even before the WMD was formed, Accra was buying too little equipment, buying equipment that was technologically too expensive and too capital-intensive, and then failing to maintain that equipment properly.[58] Service deteriorated from the mid-1970s until the creation of the WMD in the mid-1980s, improved for a while with infusions of new, fancy, imported vehicles, and then deteriorated once more into the 1990s (Cour, 1985, p. 227; Benneh et al., 1993, p. 38; WMD, 1994b, pp. 14, 25; GOG, 1994, p. 11). Now, every day is a crisis. Most vehicles regularly break down. Each day, WMD sends out what will operate to collect what it can.[59]

The solid waste that is collected is taken to one of the two Accra landfill sites, Abeka and Mallam (GOG, 1988, p. 37; GOG, 1992d, p. 144).[60] There, the refuse is minimally compacted and covered — probably the optimal degree of treatment with Accra's collection rate at its present level.[61] The City correctly ignores the World Bank's urging that "sanitary landfill practices should be adopted" (World Bank, 1989a, p. 25); it is far more important to get the refuse away from the crowded parts of the City than it is to bury it with care.

Compared to other big cities in developing countries, there is little scavenging for recyclables in Accra (as many have noticed: GOG, 1992a, p. 193; GOG, 1992c, p. 100). One occasionally sees door-to-door peddlers buying up recyclable waste products, children searching for bottles and cans, or scavengers around communal containers, but the only serious recycling goes on at the landfills.

At the landfills, no sooner are the WMD and private trucks unloaded than their contents are swarmed over by some two dozen scavengers seeking iron, aluminum, rubber, and many kinds of plastic. Only for these products is there an active market.[62] Accra's scavengers are different from those of other cities in that they are few, young, male, casual, cheerful, and well-off.[63] The explanation seems to lie in their ability to restrict entry into their scavenger-cartel. The landfill supervisor helps to enforce the cartel by being willing to call in the police whenever an "unruly" newcomer arrives.

It is hard to understand why there is so little recycling in so poor a country. Perhaps the poverty itself is the explanation — not enough trash to satisfy the minimum efficient size of recycling operations, and not enough capital and entrepreneurship to undertake such operations. Noting this private absence is *not* a call for public intervention; the landfill will still be there to be mined at a later date.

Storm drains

Accra is honeycombed with formal and informal "drains", which are filled with all the things we have been discussing — water, septage, sullage, and solid waste. They flood and overflow in the rainy season, and they become informal waste dumps or stagnant cesspools in the dry (Dorkenoo, 1995).[64]

Inadequate drainage systems such as this cause two kinds of serious social problems. One — periodic flooding — destroys property, wastes time, and causes loss of life.

And two, uncollected solid and human waste dumps in densely populated areas damage the health of the nearby residents. It is estimated that nearly a fifth of Accra's residents are subject to flooding (GOG, 1992c, p. 100); and the rate of flooding is increasing because of the location of new, unplanned residential areas and the lack of maintenance of the drainage network (GOG, 1988, p. 24f). Health effects are less obvious but surely not absent. The effects are not just localized. Most of the polluted sullage, septage, and water ends up in the Korle Lagoon, near the coast at the very center of Accra. "The lagoon is in a sorry state" — silted, unsightly, and a lost fishing resource (Akuffo, 1989, p. 19; GOG, 1994, p. 11).

To some extent, the fate of the drains derives from bad engineering and from "unsatisfactory" cleaning by the WMD, which is responsible for the drains on major streets (GOG, 1994, p. 31; GOG, 1988, p. 33). But the chief problem is that drainage systems are the ultimate in public goods — if they work, it is impossible to exclude nearby residents from their benefits, and one person's benefits do not detract from the next person's (World Bank, 1994a, p. 39). Moreover, how well the drainage system works is pretty much independent of the care given the portion of the drains in front of each house by the resident of that house.

In the absence of budget for the City to take over care of all drains and in the absence of any easy mechanism for enforcing householder responsibility for abutting drains, there will be no major change in Accra's drainage capacities in the near future. But it should be noted that a functioning system of public toilets and solid waste collection would do a great deal to improve both the health and flood aspects of the drains. Since the reverse is not true, there is reason to attach lower priority to the drains themselves.

Summary

While Ghana was the most urbanized of African countries at the time of Independence, its urban population growth rate has since fallen to the lowest in Africa. Accra has thereby been spared the problems that accompany rapid growth. But its infrastructure has suffered badly from the economic decline of the past 35 years, and the sprawling pattern of Accra's growth has meant that many new areas are inadequately serviced.

Somewhat more than half of Accra's residents receive in-house or in-yard water. Other and generally poorer residents get their water from vendors, standpipes, or neighbors; these latter sources are often expensive, and the poor as a result consume much less water than the well-off, 30-60 liters per capita per day (l/c/d) *versus* 100-150 l/c/d.

Despite the adequacy of Accra's water sources, much water is lost in transmission, and many residents suffer from intermittent flow and variable pressure. Nominal water prices have been kept low in Accra, which has meant an inadequate budget for maintenance and expansion. The resulting water problems have effectively added greatly to the real cost of water, requiring waits, walks, water tanker-trucks, and in-house water storage.

Sullage (washing water) in Accra is largely discharged directly into yards, drains, canals, and rivers. Septage (human waste) goes into septic tanks and pit or pan latrines; the sewer system covers few residents. Septic tanks are generally emptied regularly and the sludge turned into fertilizer. Nightsoil from pan and pit latrines is inefficiently collected and is simply dumped into the ocean. There are few public toilets in Accra; traditionally, these have been badly operated, although recent changes in their operation have improved them greatly.

Less than half of Accra's solid waste is collected and disposed of in landfills. Many well-off households do not subscribe to a collection service, and many poor areas are poorly serviced. Solid waste collection has deteriorated steadily in Accra, as budgetary problems have meant inadequate maintenance and replacement of equipment. There is relatively little recycling by scavengers, either in the City or at the landfills.

Accra's many drains too often become repositories for waste of all kinds, resulting in problems of health and flooding.

Notes

1   Property tax collection in Accra averaged only $1.18 per capita per year over 1988-92 (World Bank, 1994a, p. 54).

2   The sprawl has gone beyond Accra's borders into the surrounding Ga District, largely rural until 1960; today, accordingly, planning efforts focus no longer on just Accra but on the Greater Accra Metropolitan Area (GAMA).

3   Accra's water (and that of its nearby port city, Tema) comes from Lake Volta through waterworks located at Kpong and from the Densu River through waterworks located at Weija. The capacities of these two treatment plants are 172 megaliters per day (mld) and 90 mld, respectively (GOG, 1992d, pp. 120ff).

4   During these three decades, Ghana's real GDP per capita declined by nearly one half. (The metric equivalent of 30 mgd is 136 megaliters per day (mld).)

5   Some of the recent estimates of the percentage of residents with in-house or in-yard water (all the figures below refer to the entire City of Accra unless otherwise noted): in 1980, 70% (Akuffo, 1989, p. 6); for greater Accra (which includes Tema) in 1980, 61% (Tahal, 1980, p. IV-5); in Madina (a low-income neighborhood of Accra) in 1980, 64% (Ankrah, 1994, pp. 16f); and in 1991, 66% (Benneh et al., 1993, p. 11).

6   Studies of vendors vary greatly in their estimates of the price multiple, but it is at least four times as much (World Bank, 1994a, p. 25), has been estimated to be ten times as much (Bahl and Linn, 1992, p. 296; GOG, 1992b, p. 31; Stephens et al., 1994, p. 23), and may be 20 times as much — reaching 30-50 times as much during water crises (Tahal, 1980, p. III-5; Ankrah, 1994, p. 30).

7   One study of standpipe users found that, for the average family of 6.3 persons, consuming 30 liters per capita per day (l/c/d), 3.6 hours per day were required in walking and waiting for water (Tahal, 1980, p. III-7). Even at an assumed wage rate as low as $0.05 per hour, this walking and waiting time represents *five* times the monetary cost of tap water (at the lowest bracket of the GWSC tariff for in-house connections).

8   The GWSC has in the past tried three methods to ease the burden of water connections to the poor: 1) subsidizing all connections — the cost was only about $10 in the late 1970s (Tahal, 1980, p. III-8); 2) granting soft loans with well spread-out payments, but too often the loans were not

32

repaid, forcing the GWSC to cut off the water and swallow the connection cost; and 3) permitting poor families to do their own digging and plumbing, but too many illegal or leaking connections were being made, forcing the GWSC into expensive monitoring and repairs. Today, all connection work is done by the GWSC, and all payments are in advance.

9 Tahal, 1980, p. IV-4, claimed that nearly half of the public standpipes had been abandoned as part of a policy "attempt to force the installation of private connections", though the policy "in no way" succeeded in increasing the rate of new house connections (ibid., p. IV-13).

10 The GWSC tries to engage an independent agent at each such standpipe, who buys at the lowest-tariff GWSC price and sells at a higher price, earning an income on the price difference. The system seems to work, although the GWSC has had some problems with agents disappearing with their revenues and not paying for their water.

11 For an example from Ghana:

The dense and often irregular pattern of buildings ... will make it extremely difficult to construct water mains and sewers in the narrow passages and lanes between the dwellings. ... the best level of service that could be provided under the present conditions would be yard connection for 20 per cent of the houses and public standpipes for the rest. (Akuffo, 1989, pp. 6f)

12 Some of the estimates of UFW in Accra in recent years: 42-54% (Tahal, 1980, p. III-2); 45% (Engmann, 1990, Water Supply, p. 8); 42-54% (GOG, 1992d, p. 122); 58% (GOG, 1992d, p. 127); 40% (GOG, 1992c, p. 83); 67% (GWSC, 1993, p. 36); 53% (GWSC, 1993, p. 36); 20-36% (Stephens et al., 1994, p. 21); and 32% (GOG, 1994, p. 27).

13 Both water treatment works "have had working meters but these have become defective" (GWSC, 1993, p. 36).

14 Benneh et al., 1993, p. 14, and Stephens et al., 1994, p. 21, argue that something like one half of the UFW occurs in the transmission to Accra.

15 GOG, 1992d, p. 122, argues that something like two thirds of the UFW goes into illegal connections. The GWSC now plans to privatize the connection service in some cities in order to improve its revenue collection (Amoah, 1995, p. 1).

16 In 1992, for example, GWSC had to replace nearly one fifth of its meters (GWSC, 1993, p. 31).

17 In 1993, barely half of GWSC's water customers were metered. During the last few years, GWSC has undertaken "an ambitious metering

programme" and hopes to "complete" the metering in 1995 (GWSC, 1993, p. 30).

18 The government stopped subsidizing the shortfall between GWSC billings and collections in 1986, and the collection ratio has much improved as a result — from around 60% in 1986-88 to around 90% in 1989-93 (Engmann, 1990, Water Supply, p. 4).

19 GOG, 1994, p. 26, says the percentage is "over 40%."

20 96% of the households do (Benneh et al., 1993, p. 15).

21 We were unable to get exact figures for the size of this fleet, although it is apparently small — only two trucks in 1984 (Preble, 1984, p. 6).

22 To cope with supply irregularity, most high-income houses are equipped with large enclosed overhead water tanks (which cost about $0.10 per liter of capacity); the poor use open barrels, if anything (Ankrah, 1994, p. 21). When there is a water problem in an area, GWSC tankers are supposed to provide it for free, but "most of the mobile tankers are in disrepair and have been out of operation for a considerable time" (ibid., p. 25). Private tankers are readily available for the larger orders of the rich.

23 Note in Figure 2.1 that the estimated costs for interest and depreciation are small; this is almost certainly due to our serious underestimation of the capital stock in Accra's water system (Cour, 1985, p. 227). (See Appendix B for the method of estimation and the argument that capital is here underestimated.) Furthermore, the "income" of Figure 2.1 is billings, and until recently barely half of that was actually collected. In short, as careful World Bank estimates indicate, total cost, even without consideration of interest and depreciation, is surely greater than total (actual) income (World Bank, 1994a, p. 29; World Bank, 1989a, p. 11).

24 The GWSC staff should be cut by nearly one third, and there are "far too many" unskilled workers (World Bank, 1989b, p. 7). "The overall thrust of the services strategy in the short and medium term should be directed to the continued rehabilitation and improved utilization of engineering services" (GOG, 1992c, p. 82). "Continuing financial and managerial problems in GWSC remain the most serious constraints to the rehabilitation and maintenance of water supplies" (World Bank, 1993a, pp. 73f).

25 The ticks for the years in Figure 2.2 are shown at the mid-year, i.e. 1 July. For this graph, monthly CPI data were used to convert monthly prices to December 1994 prices, and then the December 1994 cedi prices were converted to U.S. dollars using that month's exchange rate. Rising block rates (i.e. lifeline pricing) were introduced in Accra in the 1970s (Bahl and Linn, 1992, p. 293).

26 These drains, designed for floods, carry not only sullage but also solid waste, and hence we leave discussion of them to a later section.

27 The seemingly greater accuracy of the 1980 figures is due simply to the fact that there was only one survey done then (Tahal, 1980). For the late 1980s, there were several, those listed above plus Benneh et al., 1993, p. 28, and Songsore, 1992, p. 7. The ranges for 1989 indicate the extent to which their findings differ.

28 Tahal, 1980, p. IV-8. Not only was there a sewer surcharge but also the water needs of a sewered household were increased — and recall, the water flows are erratic. The sewer surcharge was increased from 25% to 35% in March 1987.

29 Indeed, while the City operates 15 public toilets in the area serviced by the sewer system, very few of them are connected to it (Tahal, 1980, p. 34; GOG, 1988, p. 36).

30 For completeness, we should notice that this is the largest but not the only sewer in Accra. There are four private sewerage schemes and 14 sewage treatment plants. "Most of the plants have broken down ... [and] raw sewage passes untreated into nearby water courses" (GOG, 1992d, p. 128; Akuffo, 1989, p. 18).

31 In 1993, according to the GWSC accounts, the total sewerage revenue was over $400,000 and the total sewerage cost less than $100,000. Of course, too little was spent operating the system, and this cost figure does not count depreciation, interest, or overhead costs. Some observers have noted the need for compulsion or subsidies, or both (Otoo, 1993, p. 74).

32 Nearly three fourths of Accra's households share their toilet facilities with at least one other family (Benneh et al., 1993, p. 29).

33 The WMD was formed in 1985 from parts of the City's Health Office and Engineering Office (WMD, 1993, p. 8; WMD, 1994b, p. 91). There are no WMD data on private collection of solid waste before 1992.

34 As recently as 1988, it was estimated that 90% of the desludging of septic tanks was done by the City (GOG, 1988, p. 32). WMD data for arrivals at the sewage treatment plants in January 1995 indicate that the WMD did 64% of the desludging, large organizations did 29%, and private contractors did 7%.

35 The WMD price has roughly quadrupled (in real terms) in the last five years (Darkwa, 1990, p. 35).

36 These two treatment facilities are recent. Previously, the sludge was simply dumped into the ocean, as is nightsoil still (Akuffo, 1989, p. 21).

37 The WMD accounts for that month also charge $18 thousand to overhead and depreciation (but nothing for interest).

38 Households pay slightly less than half the capital cost of a KVIP initially and then pay the rest (interest-free) over the next two years. The charge for KVIP desludging, done entirely by the City, is also below full cost.

39 Estimates vary widely: 9,000 (WMD, 1994a, p. 4); 10,526 (GOG, 1992a, p. 101); 20,000 (Songsore, 1992, p. 6); 20,388 (GOG, 1988, p. 33); 30,000 (GOG, 1992c, p. 91); 32,350 (GOG, 1992d, p. 128); more than 33,000 (GOG, 1994, p. 10).

40 Where nightsoil collection is poor, there is added pressure on the public toilets.

41 This fee probably covers operating costs but not full costs including depreciation, overheads, and interest (WMD, 1995a, pp. 3f).

42 Some of the estimates: 200 (Songsore, 1992, p. 6); 117 (Stephens et al., 1994, p. 25); 141 (GOG, 1992d, pp. 128-130); 199 (Tahal, 1980, p. 34); 185 (WMD, 1994a, p. 4); and 155, "more than 90% ... operational" (GOG, 1988, p. 32). Our count (March 1995) is 141.

43 40% of 1,500,000 people is 600,000; 200 blocks times 40 seats each is 8,000 (Engmann, 1990, implicitly estimates an average of only twelve seats per block); 600,000 divided by 8,000 is 75 people per seat. GOG, 1992a, p. 37, surveys areas where each public toilet serves an average of 3,650 persons per day — and notes that "many residents resort to other avenues" (ibid.). Akuffo, 1989, p. 16, reports for the Nima-Mamobi (i.e. a very poor) area, usage is "400 persons per hole." Pokoo, 1994, p. 36, notes that, in poor areas, there are "long queues in the early mornings at the public toilets."

44 The charge is usually higher if it is a water-flushed toilet, if toilet paper has to be provided, or if nightsoil from pan latrines is deposited.

45 Nor is its neighboring port city, Tema — see Botchie, 1994.

46 That is, from about 0.3 (in 1988-90) out of the 0.4-0.5 m^3/c/yr generated down to 0.2 (by 1994) out of 0.4-0.5.

47 Some of the estimates of the Accra solid waste collection rate in recent years: 60% (GOG, 1988, p. 32); 75% (Engmann, 1990, Solid Waste, p. 8); 60% (GOG, 1992a, p. 2); "less than" 80% (GOG, 1992c, p. 96); 60% (Benneh et al., 1993, p. 38); 70-75% (Armah, 1993, p. 80); 80% (GOG, 1994, p. 43); 67% (World Bank, 1994a, p. 14); 43-45% (WMD, 1994b, pp. 14, 20).

48 The WMD trucks will collect from any containers but urge families to use the standard size.

49 While there seems to be no regular recycling at this stage of the solid waste flow, some families do keep out: 1) their paper to wrap market purchases, to use as toilet paper, or to burn; 2) bottles for their own use

(for cooking oil or kerosene) or for sale to occasional door-to-door buyers; or 3) scrap metal for sale when enough has accumulated to be worth the trouble.

50 Where does the solid waste of the non-participating households go? Possibly buried or burned; probably carried to the nearest communal containers.

51 On its curbside pickup program in 1994, the WMD earned revenue of $16 thousand per month, with costs of only $7 thousand per month (WMD, 1994b, p. 18). The WMD includes in this cost figure not only labor and fuel but also maintenance and administrative overheads, but it does not include depreciation or interest (inclusion of which would reduce the apparent "profit" of $9 thousand per month by at least one half).

52 Each container is supposed to serve 3,000 people, but 200 times 3,000 is barely half the City's low-income population (GOG, 1992a, p. 100). When 6,000 people use one of these containers, each adding his or her 0.5 kg/c/d to it, it needs to be emptied daily — which doesn't happen on a regular basis. Hence, they overfill, accumulating never-collected "heaps" and "informal dumps" (i.e. unauthorized collection centers).

53 For a review of earlier solid waste disposal techniques, finances, and policies, see WMD, 1994b, p. 8. Basically, communal containers had always been free until 1988.

54 On its community container pickup program in 1994, the WMD earned revenue of $4 thousand per month, with costs of $48 thousand per month (WMD, 1994b, pp. 18ff). The WMD includes in this cost figure not only labor and fuel but also maintenance and administrative overheads, but it does not include depreciation or interest (inclusion of which would increase the loss of $44 thousand per month by at least one fourth).

55 Especially in Accra, where rich and poor neighborhoods are so checkered. For example, to drive downtown from one of the wealthiest suburbs, Airport residential area, requires going right through one of the poorest, Nima.

56 Another way of looking at this difference: the WMD averages a revenue of $5.00/m^3 on the solid waste it collects at curbside, while the private contractors average $3.65/m^3 (WMD, 1994b, pp. 58f).

57 These private sector operations leave "much room for improvement" because either they try to charge for communal containers, which leads to "indiscriminate dumping," or they leave the WMD "to provide parallel communal container refuse collection services" (WMD, 1994b, pp. 10ff).

58 The WMD itself has estimated that the tractor-trailer costs only 28% as much per household serviced as a compactor truck (WMD, 1994b, p. 74).

59 At the nadir, in 1985, of the 55 solid-waste vehicles in Accra, "only 6 are roadworthy" (Oddoye, 1985, p. 39). In 1990, it was claimed that 80% of the vehicles were operating (Engmann, 1990, Solid Waste, p. 8). Today, barely half (World Bank, 1994a, p. 35).

60 The private contractors pay a modest tipping fee there ($2.50 for pickup trucks and $5.00 for full-size tippers and trucks). A small portion of Accra's waste goes to the Teshie Compost Plant (Armah, 1993, p. 81).

61 The Mallam site is a traditional dump, with much dust, litter, fire, and smoke. For some reason, the Ghana Environmental Protection Council has started monitoring the groundwater, almost surely prematurely (Benneh et al., 1993, p. 43).

62 There was some collecting of glass — broken, since otherwise bottles are not discarded — which is resold for use as the topping for security fences. There was some collecting of paper, not to be recycled into new paper but rather for use as wrappings for food products sold in the market. Indeed, because he thought this practice unhealthy, the landfill supervisor has forbidden paper collection, and the scavengers seemed to have accepted that edict.

63 While talking to one scavenger who was taking the afternoon off in his small porched shack, sitting in a wheelbarrow lined with pillows and working on a Rubik's cube, we learned about a lucky colleague who recently found an envelope with over a thousand U.S. dollars in it — the colleague was "on vacation."

64 Coincidentally, Accra's worst flooding in a half century occurred this year, killing "more than 20" people and causing millions of U.S. dollars in property damage (Abugri, 1995, p. 5).

3 Harare

with Albert Mafusire

Introduction

The Province of Harare, which consists almost entirely of the City of Harare and its dormitory suburb, Chitungwiza, contained a million and a half residents in the 1992 Census (CSO, 1994, p. 13). Harare and Chitungwiza together have grown at a rate of over 5% per annum since Rhodesia's Unilateral Declaration of Independence (UDI) in 1965 and at a rate of nearly 6% per annum since Zimbabwe's Independence in 1980; currently, they are growing at 6-8% per annum (COH, Master, 1993, p. 8; Mubvami and Korsaeth, 1995, p. 4). These growth rates and the resulting degree of urbanization are very comparable to those of other Sub-Saharan African capital cities (World Bank, 1989c, p. 1).

The comparability of Harare to other African cities, however, ends there. Harare is an "apartheid city" (Porter, 1993b). It was constructed and inhabited by both blacks and whites over a century ago, and named Salisbury; but it was operated *by and for* whites until Independence, when it was renamed Harare. What does it mean to operate a city for the benefit of a portion of its residents? As an official urban plan stated:

> The present purpose of Salisbury is to provide a means of livelihood as well as security, social and cultural amenities for its *inhabitants* appropriate to the way of life of its *citizens*. (Quoted in van Hoffen, 1975, p. 148; emphases added by van Hoffen)

An apartheid city, like the entire apartheid system, was the white government's attempt to resolve the dilemma that the exploitation of blacks

39

requires their proximity to whites even though such proximity poses risks to the process of exploitation (Porter, 1978a and 1984).

In Salisbury, whites lived mostly in spacious houses and grounds sprawling out toward the north of the central business district (CBD). Black housing, on the other hand, was located in dense patches "a considerable distance from the city centre," mostly to the south and west, and originally consisted of hostels for single male adults (Zindere, 1991, p. 21).[1] By the 1970s, it had proven impossible to prevent black workers from bringing their wives and families to Salisbury, and conventional low-cost family housing began to appear in the black townships.[2] Also in the 1970s, the ultimate cruelty of apartheid saw the rise of Chitungwiza, a densely populated black dormitory city consisting of row on row of identical "core houses" and "ultra low cost houses," located 25 kilometers (km) to the south of Salisbury — with nearly one hour's bus commute each way (Rakodi and Mutizwa-Mangiza, 1989, p. 10; Zindere, 1991, pp. 21ff; Rambanapasi, 1994, p. 208).

The black townships of Rhodesia were always well serviced. The hostels had water and toilets, and the two or three bedroom houses had "a relatively high standard of utilities" (Potts, 1994, p. 210). Because of these high standards, together with low black wages, it was necessary to subsidize the houses, and this meant that the City's budget generally ran out before the demand was satisfied. In the 1970s, the white government tried to solve the problem by reducing the housing standards. In fact, however, the problem became worse as "influx controls" broke down and blacks flooded into Salisbury in the late 1970s to escape the rural havoc of the war for Independence.[3]

At Independence, the low-income housing (i.e. black housing) of Salisbury-becoming-Harare was "severely overcrowded" and squatter areas "were expanding" (Mbizi, 1990, p. 9). Those who owned houses were renting out rooms partly because the City's charges on owners were high and partly because the unsatisfied demand of those who could not acquire new houses kept rents high. Not only were rooms rented but also densely packed "sheds" were springing up in the yards.

Thus, the new government at Independence inherited both a shortage of housing for the low-income residents of Harare and a standard of housing that was unaffordable by most low-income residents. Being unwilling to lower the standards below the pre-Independence level, and being constrained by budget limitations, the new government could only afford to build at a slow pace, and the housing shortage worsened. Today, roughly half the households in Harare and Chitungwiza are listed as "lodgers" (CSO, 1994, p. 89). Ironically, the post-Independence government, determined that houses should be of high

40

standards or not exist at all, continued the pre-Independence assault on squatter settlements — this "severe anti-squatting stance ... is probably the most stringently enforced in Africa" (Potts, 1994, pp. 211ff; Patel, 1988, pp. 209ff; Rakodi and Mutizwa-Mangiza, 1990, pp. 18f).

Thus, Independence for Zimbabwe has not meant an end to the colonial idea of two cities — one for wealthy whites and one for poor blacks — but rather a perpetuation of it, although many high-income and middle-income blacks have taken advantage of the end of official racial segregation and the subsequent white exodus to escape the high-density living of the former black townships.[4] Today, roughly two thirds of the people of Harare live in what are now called "high-density residential areas" and barely one tenth in "low-density residential areas" (COH, Master, 1993, p. 10). The dramatic difference in densities is shown in Table 3.1.

Drinking water

Nearly all of the households of Harare and Chitungwiza receive piped, treated water, either inside their houses or from yard-taps directly outside (Table 3.2). But this amazing achievement by developing country standards is not as praiseworthy as it at first seems, for it is achieved almost by definition. All officially recognized residential areas of Harare, even those of new development, have access to piped water; and all officially recognized properties, even in high-density, low-income areas, have their own metered water supply (Colquhoun, 1990b, p. 5.10). In short, all officially recognized housing in Harare has running water — housing without it is just not recognized.[5]

Residents of "unofficial" areas rely for their water on shallow (often contaminated) wells or taps attached to public toilets. The City does not provide public standpipes. While it sometimes drills boreholes, many of these are dry or without functioning pumps (Ziracha, 1989, p. 34; Mbizi, 1990, p. 95; Butcher, 1993, p. 71); the City does not charge for this borehole water (Potts, 1994, p. 215).

Statistics on water consumption, therefore, apply almost entirely to the "officially recognized" areas. Overall, in Harare and Chitungwiza, water production (i.e. the output of treated water) is about 250 liters per capita per day (l/c/d) — see Figure 3.1 — but much of this goes to public, commercial, and industrial uses. Residential consumption is generally lower, sometimes much lower. Indeed, water consumption in Chitungwiza averages only 47 l/c/d (Rambanapasi, 1994, p. 218). As shown in Table 3.3, people living in the

41

Table 3.1
Housing densities in different parts of Harare

Residential Area	Persons/Hectare	Houses/Hectare
Low-Density	9.24	1.24
High-Density	109.46	11.20

Note: Data for 1987.

Source: Zinyama, 1993, p. 16.

Table 3.2
Type of delivery of Harare drinking water

Percent of residents with piped water

Residential Area	Inside	Outside	Neither
Urban Harare	28.5%	62.1%	9.4%
Chitungwiza	44.6%	54.6%	0.8%
Overall	31.2%	60.5%	8.3%

Note: Rural Harare is also included in the "overall" figures — it is a Census area consisting of white farms and black squatter settlements, in total less than 2% of Harare Province's 1.5 million residents.

Source: CSO, 1994, pp. 91, 115.

Table 3.3
Harare water consumption by type of housing

Residential liters per capita per day (l/c/d)

Area	1982	1992
High-Density	86	73
Medium-Density	200	200
Low-Density	300	311

Source: COH, Master, 1993, p. 30.

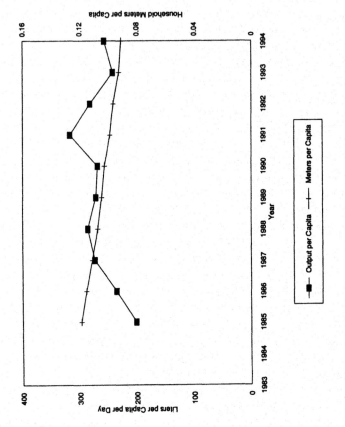

Figure 3.1. Water production and meters (in 1/c/d and meters/pop.)

44

high-density (i.e. low-income) areas get much less water than those living in medium-density or low-density areas; and more importantly, the rich appear to be getting more water (per person per day) than they did a decade ago and the poor appear to be getting less. There are two reasons for this decline in water consumption per capita by the poor of Harare.

The first reason is the crowding of the residences in low-income areas. As more and more people share a water tap, each receives less access to the tap than before. There is both increased queueing and decreased privacy involved with each water use.

The second reason is the low pressure in the water system due to the heavy demands being placed on it in the high-density areas — the pipes are not large enough to satisfy these demands.[6] Low pressure is a form of high price, and it reduces demand — indeed, in some developing-country cities, the water pressure has been reduced intentionally as a conscious water-rationing device.

One might think of price changes as another reason for altered demand. However, as Figure 3.2 shows, the real price of water in Harare — while saw-toothing from tariff change to tariff change — has shown little trend over the past decade.[7] For those households consuming 10 cubic meters per month (m^3/mo), the real cost per month actually fell by one third (i.e. from $1.18 to $0.80) between July 1984 and December 1994; and for those households consuming 30 m^3/mo, the real cost per month fell by one fifth (i.e. from $4.56 to $3.58) over that same period.[8]

With increased numbers of people in each residence, however, the number of people on each meter has risen, so that families consuming the same amount of water *per person* as before will nevertheless be consuming more water *per meter*; as a result, they may be moving into higher water-tariff brackets (as indicated by the trends of Figure 3.1). To illustrate with the numbers above, if a single household with a single meter tripled in size between 1984 and 1994 and its total consumption rose from 10 m^3/mo to 30 m^3/mo, its household water bill would have risen from $1.18 to $3.58 — and the real water price *per capita* would not have changed at all.[9]

This failure of the real price to rise over the past 15 years is surprising, considering the concern shown by the Harare City Council and Department of Works about the availability of water. There are four kinds of evidence of great scarcity — and hence of costliness — of water in Harare:

1   Hardly a week goes by without a newspaper report of the threatening water situation — four, for example, in one month in one newspaper, *The Herald*.[10]

45

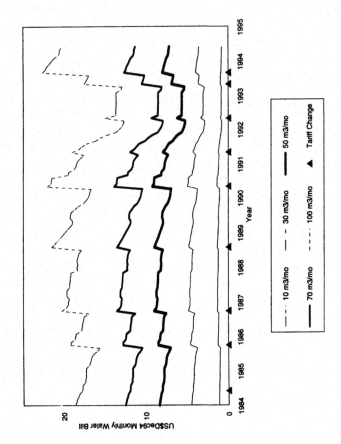

Figure 3.2. Monthly water bills, 1984-95 (in US$ Dec 94 per month)

2   Harder evidence of the increasing costliness of meeting Harare's water demands is seen in Figure 3.3.[11] The production of treated water in Harare has grown very slowly compared to the real investments made — it has taken almost \$1 of investment to produce each cubic meter of new water per annum.

3   The past and present willingness of the City Council to invest in expensive sewage treatment facilities in order to recapture some of the treated effluent for use as drinking water hints at the real opportunity cost of water.[12]

4   Conversations with officials in the Department of Works reinforce the idea that Harare's water prices fall far short of covering even the real operating costs of providing the water. Their guess is that the marginal variable cost of providing water is now about $0.50/m^3$, and a crude application of the data of Figure 3.3 suggests that "long-run marginal cost" (i.e. average total cost) is around $0.60/m^3$.[13] On average, Harare's water users pay only about $0.15/m^3$, *less than one fourth* of the real marginal cost of the water.[14]

Water pricing in Harare is highly politicized, and neither the City Council nor the Central Government (which must approve all water tariff changes in the high-density residential areas) wants to raise the price of water. Both appear quite satisfied just to cover the current, nominal, accounting costs. The water tariffs do achieve this (see Figure 3.4) — both water income per capita and water expenditure per capita have remained steady at around \$12 per annum over the last decade. But, if we search for an estimate of average total cost by including interest and depreciation on the water capital stock, expenditure per capita is nearly doubled.[15]

The World Bank describes the process euphemistically: the City's water accounts

> are funded by tariffs ... levied by the City Council and are intended to cover the full cost of these services ...; the tariffs are generally revised as necessary to keep abreast of changes in costs." (World Bank, 1989c, p. 41)

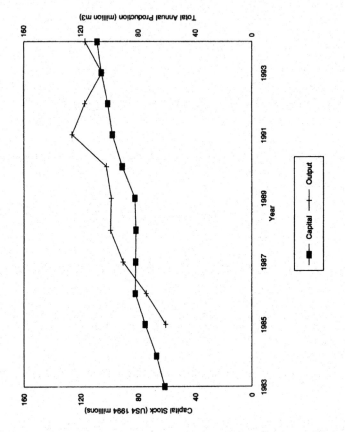

Figure 3.3. Water capital and output (in US$ 1994 and cubic meters)

48

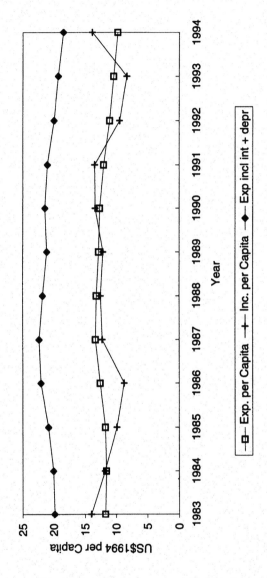

Figure 3.4. Water income, expenditure (in US$ 1994 per capita)

49

But Mubvami and Korsaeth put it more accurately:

> ... the tariff system bears no relationship to the cost of the production of the water. Harare bases its costs on anticipated annual expenditure rather than the cost of delivering a unit of water to the users. (Mubvami and Korsaeth, 1995, p. 15)

We cannot excuse Harare's low water price by simply calling the pricing process political. It is politically difficult to raise water prices because so many poor people have piped water and flush toilets, both of which encourage more water use by the poor than they can really afford at realistic prices. The more fundamental question is why the poor are forced to acquire a water and toilet system that would be beyond their means if it were priced at the social cost of its provision. It was politically tempting, after the end of apartheid, for the new government to insist that everyone's environmental services be raised to the level that only the whites could previously afford, but as a practical matter such insistence is costly — and the costs are borne by foregone investment opportunities elsewhere. In Harare, those foregone investments include, among others, investments in new water supply and in new housing. For many poor people, a completed house with nearby standpipe and public toilet might be preferred to the present water-and-toilet-first plot (provided in-house water and sanitation could be later added).

The municipal piped water supply is not the only water that is artificially cheap in Harare. Borehole water is free, in the private sense that, once one has invested in the drilling, there is no variable cost to removing the water.[16] The City neither restricts the drilling of boreholes nor places meters or charges on the pumping out of groundwater.[17]

Essentially the City has lost sight of the fact that water is water, and this is what is scarce. This fact is often obscured by the many ways in which surface water and borehole water differ — they differ in chemical details, in flavor, in rates of recharge, in treatment cost, in delivery cost, and in who pays for the delivery (i.e. the capital costs and the pumping). But the two kinds of water are to a great extent interchangeable, and they should be lumped together when thinking about Harare's total water potential.

There are two potential problems with Harare's laissez-faire attitude toward borehole water use.

One, it invites the depletion of underground water, which can be a permanent and irreversible loss. If such water is tended to properly, it might prove to be a renewable and perpetual water resource for Harare. Little is known about the groundwater situation under the City, but there are hints that it is not inexhaustible. New boreholes need to be drilled deeper than they used to be, and flows from existing boreholes are not as good as they once were.

And two, the absence of a price on groundwater invites owners of boreholes to think of groundwater as a free good, to be used freely and even frivolously. Indeed, every time the City discovers its water to be scarce and raises the price of piped, treated water, it increases the incentive to borehole-owners to use even more groundwater.

This two-price system for water leads to all kinds of inefficiencies. For example, in March 1995, the City Council forbid the use of piped water for garden irrigation, but at the same time encouraged the use of groundwater for this same purpose (*The Herald*, 1995d, p. 4). Water is water. Either it is efficient to utilize water for market gardening, or it is inefficient to waste water in this way. It should not matter which water source is involved.

The City should claim ownership to the groundwater beneath it, requiring licenses from those who drill boreholes, metering the water that is extracted from these boreholes, and charging for that water. The proper charge for borehole water may be higher or lower than the charge for piped water.[18] In Harare, however, there are three reasons why the City should charge less for borehole water than for piped water: 1) as an interim measure, the City might wish to soften the shock of introducing charges on what was previously treated as free; 2) the City does not have to treat and distribute privately drilled borehole water, and hence the marginal cost is lower for such water; and 3) the groundwater may be sufficiently plentiful, and free of danger of depletion, that the City may want to encourage the use of groundwater in place of surface water, especially when surface water is in low supply. But all three of these reasons suggest only that the City price for groundwater be lower than the price for surface water, not that groundwater should always have a zero price.

[The previous seven paragraphs have been left largely unchanged since the first draft of this monograph was circulated. But some good news has emerged since then. The Minister of Lands and Water Resources announced on 18 October 1995 that "new water pricing policies are being investigated .... the new prices will cover underground and surface water." Unfortunately, some uses of borehole water, namely "domestic consumption for people and livestock," would remain exempt from such charges. At this same time, the Department of Water Resources indicated that, in the future, "one would need to apply for a water right in order to drill a borehole" although here, too, such permission would remain unnecessary if the water were intended for people or livestock. Finally, at this meeting, "higher and more realistic water prices for all users" were announced (all quotes in this paragraph from *The Herald*, 1995f, p. 1).]

Increased water revenues from higher piped water prices and the introduction of borehole water prices are necessary for greater self-financed investment in water supplies because there is probably not much scope for reducing water expenditure. The system is considered efficiently operated; water losses are low, bill collection rates high, and illegal connections minimal (COH, Master, 1993, p. 28; COH, Annual, various years; COH, City, various years; Colquhoun, 1990a).

Human waste

In no respect is Harare's "high standard of utility provision" more compelling than in its treatment of human waste (Rakodi and Mutizwa-Mangiza, 1990, p. 18). Almost all who live in Harare's recognized residential areas — high-density areas as well as low-density areas — now have access to water-flushed toilets.[19] And almost all who need to be attached to sewers are attached — though many in the high-income, low-density residential areas, with 2,000 m^2 (i.e. about one half acre) or larger plots, are permitted and have preferred septic tanks.[20]

Partly, this high standard of septage disposal was attained before Independence, as the apartheid government had long insisted on adequate water and sewage facilities in the high-density "townships" (Potts, 1994, p. 210). After Independence, however, many of the existing high-density areas were upgraded from pit latrines to water-flushed toilets and from septic tanks to sewer attachments, and all planned new high-density housing developments have been sited with access to sewer mains — alternative methods of sanitation "are not used" (Colquhoun, 1990b, p. 5.11).

While this is an achievement matched by no other city with so low a GDP per capita, the Harare situation is nevertheless not without problems.

The first problem is that Harare's *unplanned* new settlements have to make do with much less in the way of sanitation facilities. At best, the residents of such unplanned settlements will have access to an uncongested and well cleaned, ventilated, improved pit latrine — called a Blair toilet after the Harare research center at which it was developed (Colquhoun, 1990b, p. 5.11).[21] But many of the residents have to make do with a heavily shared, seldom cleaned, badly constructed, and/or unimproved pit latrine, often sited near a shallow well that is used as a water source (Ziracha, 1989, p. 35).

In one low-income area (Epworth), for example, the government has sunk many wells, subsidized Blair toilets, and offered sewer attachments to those who build up-to-standard housing (Morgan, 1987, pp. 56ff; Potts, 1994, p. 215). Nevertheless, some 80% of the residents are "still using unimproved pit toilets or open spaces" (Butcher, 1993, p. 71). And some people there continue "building substandard houses and renting them out instead of building proper houses for themselves", which means that

> water and sewage pipes which were installed in the area have been lying idle. They can only be installed where standard houses have been built. (*The Herald*, 1995b, p. 5)

In the newer shantytown areas, the City now provides free public toilets and collects nightsoil, but the facilities are badly overused and too infrequently emptied and cleaned.[22]

The second problem with Harare's handling of septage emerges in the planned settlements, where water-flushed toilets are provided to all. The problem is that "the costs of providing this service are considerable" (Mafico, 1991, p. 113). For many of the poor who are required to acquire water-flushed toilets, less luxurious facilities would be preferred. To some extent, when externalities are involved, it is quite sensible for a government to override people's private preferences; and externalities with respect to both health and groundwater certainly are generated by sewage.

On the other hand, such public insistence on better toilet facilities than low-income households want either 1) requires subsidies from the government or 2) lowers the standard of living of the poorest citizens. Harare has had to utilize some of both of these methods, for the difference in the costs of water-flushed toilets and Blair toilets is very large. The capital cost of a water-flushed system is around $520, and the present value of the operating cost is another $610; the present value of both capital and operating costs for a Blair pit latrine is only $370, about one third as much (de Kruijff, 1981, pp. 29f, 51; Mafico, 1991, p. 113).[23]

The City seeks in two ways to soften the impact of this high cost of water-flushed toilets on the poor by the way that it assesses sewer charges. The first way is that residential sewer charges are kept low by setting a high fee on commercial users to cross-subsidize residential use.

The second way that the City seeks to make its sewer fees less onerous is by not charging the full costs of its sewerage operations. Figure 3.5 shows the City's income from sewer fees and its out-of-pocket expenditures. They are roughly equal, but the expenditure figures do not include anything for

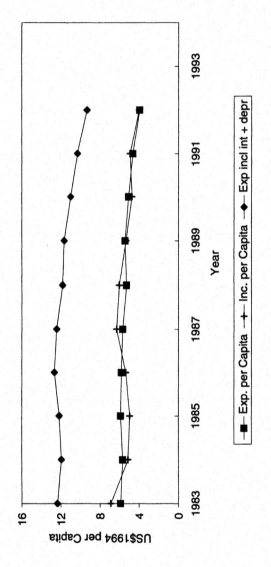

Figure 3.5. Sewerage income, expenditure (in US$ 1994 per capita)

54

depreciation or interest on the capital involved. Adding in a rough estimate of those costs suggests that total cost per capita is twice as high as the City thinks, and hence that a sizeable sewerage deficit is being run, from the viewpoint of the real, economic costs involved.[24]

Unfortunately, there is a third way in which the high cost of flushed toilets is being partly covered — by the reduced well-being of those who are forced to pay more than they are willing to pay for these services. One aspect of this reduced well-being should be noticed here. People may react to the high cost of housing, brought about by an overly high sewerage standard, in any of three ways:

1   by renting housing rather than building it, often from absentee landlords who have been known to rent as a room what should have been the house's toilet;

2   by consuming smaller housing, which in turn means greater population density per square meter (m^2) in low-income areas, hence more toilets per m^2 and hence in the end — ironically — an overloading of the sewers;[25] or

3   by constructing their houses more slowly, which means living for long periods in "toilet towns" (as they are called in Harare) because the water-and-sanitation standards have left them unable to finish their houses expeditiously (Potts, 1994, p. 214).

Since water-flushed toilets in every house are the ultimate goal of development, these reservations about premature timing are *not* a suggestion that flush toilets should be removed. Rather, in the future, in very low-income neighborhoods, people should be given greater choice about the sanitation system they want. Whatever is offered should be upgradable into flush toilets at some future time but should be lower-cost than flush toilets at the present time. Either individual Blair toilets (or twin pit latrines) or communal public toilet blocks are the obvious alternatives to be made available — which of the two depends greatly on the anticipated density of the residential area (de Kruijff, 1981, p. 37; Mafico, 1991, pp. 113f).[26]

Harare is also unusual among cities in developing countries in the degree to which its sewage is treated. All of Harare's five sewage treatment plants undertake primary treatment of the sewage, and "at least half" of the City's wastewater receives more advanced treatment (*The Herald*, 1995c, p. 6; MLAWD, 1993, p. 1). In fact, however, none of this treated water is used as

55

drinking water. After receiving "very expensive but efficient advanced wastewater treatment," the effluent was free of pathogenic bacteria but was too high in nitrates and phosphates and was commingled with untreated storm water and "poorly treated" sewage from Chitungwiza (Masundire, 1994, pp. 148ff). The effluent is currently diverted to municipal farms for irrigation of pastures and crops.

The question nags whether advanced sewage treatment is worth the cost. During the last decade, Harare has spent more on its sewage treatment than it has on its sewerage reticulation systems, and it is not clear that it is getting enough benefit from this to warrant the costs. Apparently, no benefit-cost calculations of any kind preceded the City's recent decisions to move toward more advanced sewage treatment. Even if it is true, as City officials maintain, that the costs are coming down and that the central government forces the City to undertake this advanced treatment, such benefit-cost calculations are still appropriate.

Solid waste

The City of Harare estimates that its residents generate some 0.5-0.6 kilograms per capita per day (kg/c/d) of solid waste, and the City collects and disposes of some 90-95% of this waste (Tevera, 1993, p. 86). Not all observers agree that the collection rate is this high, but visual inspection quickly informs any observer that Harare does an excellent job relative to other cities at this level of development. Partly, the priority handling of solid waste stems from its pre-Independence days, when Salisbury prided itself on being a "garden city"; but the post-Independence government has also made a strong effort to prevent deterioration of this service.

The Amenities Division of the Department of Works picks up residential trash at curbside at least once a week and provides, at that time, a replacement plastic bag free. The service is regular and reliable, and the trucks will usually pick up refuse other than, and in excess of, the plastic bag.[27]

Whether a residential area is serviced once or twice a week is "historically determined", which means that the low-density, high-income (formerly all-white) areas get the semi-weekly service and the high-density, low-income areas get the weekly service.[28] This may have made sense to the white apartheid government, which intended that whites get the lion's share of the urban amenities (Porter, 1993b); but it makes no sense today since there is much more trash being generated per square meter (though not per person) in the high-density areas, and the trash there contains a larger proportion of putrescible matter. The City is considering moving to once-weekly service in all residential areas, but more from a shortage of refuse vehicles than from an eye to equity or efficiency. Thus, it may be making the right move, even if for the wrong reason — only at a much higher standard of living than Harare residents now enjoy does twice-weekly refuse pickup begin to make willingness-to-pay sense.

The adequacy, or more precisely the inadequacy, of the number of operating solid waste vehicles has always been a problem in Harare. The number of operating vehicles declined steadily during the 1980s, and the City averaged only 35 operating vehicles by 1991:

> ... the shrinking and deteriorating vehicle fleets ... have degraded the quality of solid waste collection services. (World Bank, 1989c, p. 28)

In the last few years, the number of trucks has been greatly expanded, reaching 55 in 1993 (COH, Annual, 1994, p. 130). The newer vehicles are modern compactor trucks, but most of the vehicles are old front-loaded lift-trucks and high-sided open-top trucks, sensibly simple and labor-intensive, but now requiring frequent maintenance because of their age. Further, a largely unplanned equipment acquisition process has left the City with a very heterogeneous truck pool, causing problems of know-how and spare parts in maintenance. Rarely are as many as half of the vehicles operating on any given day.

Solid waste is also picked up by street-cleaners, over 600 men and women who ply Harare's streets with a barrel on wheels and a broom.[29] Each street-cleaner manicures but a short stretch of road, which accounts for the very neat appearance of the main streets of Harare. Indeed, the City's expenditure on street-cleaning *exceeds* its expenditure on solid waste collection and disposal — see Figure 3.6.[30] Although such labor-intensive activities as street-cleaning are politically tempting in any city with high and visible unemployment, one cannot resist thinking that a few more mechanics and spare parts in solid waste vehicle maintenance would be a more socially beneficial allocation of budget.

With the urging of the World Bank, Harare intends that solid waste collection "be self sustaining financially" (COH, Annual, 1994, p. 132; World Bank, 1989c, p. 41). To this end, "refuse fees" are levied — originally, these charges were added every six months to each property-owner's "rates" (i.e. property tax), but now they are levied separately and monthly. The real level of these solid waste fees is shown in Figure 3.7, for households that receive pickup service once or twice a week.[31]

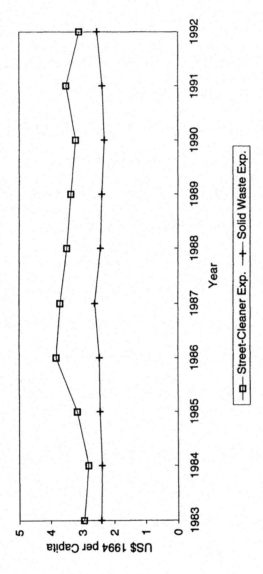

Figure 3.6. Waste disposal expenditure (in US$ 1994 per capita)

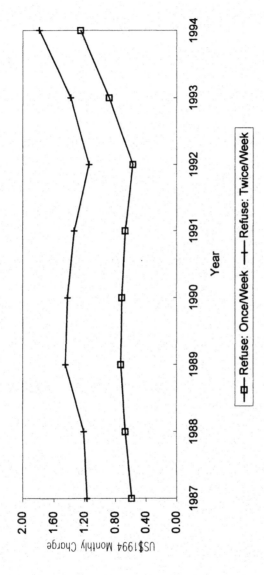

Figure 3.7. Monthly solid waste fees (in US$ 1994 per household per month)

Over the past decade, the refuse levy has cost the average resident of Harare about $2 per capita per year. But the levy does not in any way alter the *marginal* private cost of a person's use of the City's solid waste disposal system, and hence the levy in no way deters people from burdening the disposal process. To the extent that this fee is intended to encourage source reduction of waste — as a sort of Pigovian tax on a negative externality — then it fails totally.

If the purpose of the refuse levy is to generate enough revenue to cover the cost of the solid waste disposal system, it again fails. It does cover the operating cost of the system — also about $2 per capita per year over the past decade — but it adds nothing toward depreciation or interest. If these are included, then the cost per capita per year is nearly $3.[32]

The natural question, then, is just what is this refuse levy? The answer is that it is nothing more, nothing less, than a regressive addition to the property tax. It adds a few dollars a year to each household's property tax — the same amount for rich and for poor — under a misguided belief that each revenue source in the general fund should be earmarked to an expenditure destination:

> Refuse collection, disposal sites and septic tank fees have this fiscal year increased by approximately 33% indicating that waste management can be self sustaining financially .... (COH, Annual, 1994, p. 132)

One might argue that the rich get twice a week service, that they do not generate twice as much trash, and hence that they do not cost the City twice as much, while they do pay twice the refuse levy.[33] But the City began charging less than double for twice-weekly service in 1993, for reasons that have never been stated; and, if the City goes to once-weekly service everywhere, it will presumably charge all residents the same fee; so any vestige of cross-subsidization will be soon lost.

Once the solid waste is collected, it is taken to one of two dumps, the City owned Golden Quarry Municipal Garbage Dump, five kilometers to the west of the CBD, or a privately owned dump at Teviotdale, ten kilometers to the north of the CBD.[34] The Teviotdale site is free to the City, as its owner wants to fill the abandoned quarry there, but its greater distance makes its use cost-effective only for the collection vehicles servicing the northern (low-density) suburbs. As a result, some 85% of Harare's solid waste goes to Golden Quarry (COH, Annual, 1994, p. 131).

Golden Quarry is owned by the City, and it was operated by the City with leased equipment until last year. Now, a private firm does the landfill under a contract that costs the City less than its previous equipment rentals. It is a huge site, apparently with many years of life left.[35] Beside the Department of Works trucks, private haulers use the site, at a tipping fee of $5-10 (depending on the vehicle size). It is a true "dump" with frequent fires, much dust and litter, little and slow covering, liquid waste separate but in open trenches, and no environmental monitoring of groundwater. Harare, correctly we believe, spends its resources getting the waste out of the populated areas, not in burying it carefully.

As in other developing country cities, Harare's solid waste is some two fifths vegetable-putrescible matter and another fifth inert matter (i.e. ash from cookfires and soil from yard-sweepings), so the potential for recycling is limited.[36] But scavenging for recyclables is a low-skill occupation, with comparatively free entry, demanding little capital, and attracting many workers in Harare as in most other poor and labor-surplus countries. There is scavenging throughout the City, by house-to-house dealers and by workers in large businesses as a wage supplement; but much of Harare's recycling activity takes place at the landfills.

At the Golden Quarry landfill, as the refuse trucks spill their contents onto the ground for the bulldozer to spread and compact, the scavengers attack — some 200 of them, about equal numbers of men and women (with no children in sight).[37] In principle, the scavengers are the employees of recycling corporations, and each scavenger is supposed to be hired, supervised, and licensed by the parent company.[38] In fact, only about 20 or 30 of the scavengers at Golden Quarry are properly licensed.

Each scavenger specializes, scrabbling for his or her materials, and piling them up off to the side (Tevera, 1993, p. 88). The scavengers are principally seeking paper, cardboard, plastics, and household durables that are capable of reuse.[39] In the lulls between vehicle arrivals, the scavengers clean, neaten, and tie up their bundles. The household durables find their way to second-hand markets in Mbare and other high-density residential areas; the rest is sold directly to recycling companies.[40]

Authorized recyclers — who have won bids with the City for this right — send trucks periodically to the landfill, at which time the scavengers' bundles are inspected and weighed, with a hanging scale on a pole held up by two scavengers.[41] The weight is recorded against the scavenger's name or number, and payment is made at the end of the week. Scavengers earn an average of about $6 per week — about $1 higher than the minimum wage for domestic workers — in a job that requires hard, unpleasant effort and imposes substantial health and injury hazards, although there is always the possibility of "striking it rich" (Tevera, 1993, pp. 87ff).[42]

The recycling companies are many, are large, are old, and recycle a wide scope of paper and plastic materials. This in itself is surprising in a country of Zimbabwe's population size, population density, and GDP per capita (i.e. all relatively low). To know why unsubsidized recycling is so advanced might help other countries to initiate or expand their recycling industries, but there are at this time only conjectures: Zimbabwe's forest laws make virgin pulp expensive; the world's sanctions against Rhodesia during the UDI period made the import of virgin materials difficult; the import licensing regimes of the 1980s made foreign supplies again uncertain and/or expensive. The problem with all these conjectures is that the recycling firms were mostly established in the 1960s.

There are proposals afoot in Harare to reorganize this whole recycling process, with franchised recyclers operating within the City itself. The recent history of West Europe and North America informs us that this will ultimately be necessary, but the current recycling system is very labor-intensive and hence probably much more cost-effective at this stage of Zimbabwe's development.

Summary

Harare is an "apartheid city," constructed and operated until 1980 for the convenience and well-being of its white residents. As a result, at Independence, the white residential areas were spacious and well-serviced, while the black residential areas were dense, distant, and less well provided with the urban environmental amenities. By apartheid standards, the black townships of Salisbury/Harare were comparatively well serviced, but the havoc of the war for Independence and the concomitant breakdown of "influx controls" left them extremely crowded. Efforts since Independence have focused on closing this amenity gap, but the City's rapid growth and budgetary problems have slowed the effort.

Nearly all households of Harare and Chitungwiza receive piped, treated water, either in-house or in-yard, but water consumption varies greatly by location. Consumption is 50-70 liters per capita per day (l/c/d) in the high-density areas and 200-300 l/c/d in the medium-density and low-density areas.

Real water prices have not been raised over the past decade, despite evidence that water is becoming increasingly scarce and costly to the City. Water revenues cover current costs, but they do not cover full costs (including depreciation and interest), and full costs in the long run are expected to rise. Curiously, considering the expected shortages, borehole owners are permitted to use their water freely, in the sense that the City places neither restriction nor charge on such water.

Almost all residents have access to water-flushed toilets and attachment to the sewer system. The costs of this provision are high, and in the poor areas these facilities are provided by means of government subsidies and reduced levels of other, perhaps preferred, amenities. Here, too, fees are kept low, but the result is that the fee revenues only cover the operating costs.

Solid waste is picked up once or twice a week throughout the City. The City has made a strong effort to apply refuse levies to make the solid waste collection "self-sustaining financially." Recycling is highly developed in Harare, with authorized recyclers bidding for the rights to the landfill recyclables and the scavengers themselves technically employees of these recyclers.

Notes

1   The single-male dwellings emerged as a partial resolution to the dilemma that blacks were not considered permanent residents of apartheid city — but rather "temporary sojourners" — even though they were needed permanently as a labor force there (Rambanapasi, 1994, p. 199).
2   For a fuller review of housing structures, styles, standards, and services in Salisbury/Harare, see Rakodi and Mutizwa-Mangiza, 1989 and 1990; Colquhoun, 1990b; Teedon, 1990; Mbizi, 1990; Zindere, 1991; Wekwete, 1994; and Potts, 1994.
3   "Influx control" was apartheid's means of preventing blacks from entering the cities until they were needed as workers. In Rhodesia, The African (Urban Areas) Accommodation and Registration Act of 1951 required all urban blacks to register; The Vagrancy Act permitted urban authorities to expel any blacks who were unregistered or unemployed; and The Land Tenure Act specified where blacks could live and encouraged the destruction of squatter settlements. As a result of these policies, the proportion of blacks living in Rhodesia's cities did not increase until after 1975, when the escalating war reduced the government's ability to control the immigration (Seager, 1977, pp. 83f).
4   See Wekwete, 1994. Cumming, 1993, in her examination of the names on electric bills in Mabelreign, found that the percentage of blacks in Mabelreign — before Independence, a middle-income all-white area — rose from zero at the end of The Land Tenure Act in 1978 to between 23% and 87% (in different sub-areas) by 1987.

5   Not only running water but also sewerage reticulation, electricity, and access roads are required (Zindere, 1991, p. 5). Even serviced sites, on which the owner builds the rest of the house, always have installed water and sewerage systems (ibid., pp. 20ff). See also Potts, 1994, p. 210.

6   Most of Harare's water mains were installed soon after World War II, but some are as much as 80 years old. Pipes are being replaced, gradually, but the city does not have a regular programme of maintenance and rehabilitation of the network.... maintenance is carried out on a crisis basis. (Mubvami and Korsaeth, 1995, pp. 5, 15)

7   The ticks for the years in Figure 3.2 are shown at the mid-year, i.e. 1 July. For this graph, monthly CPI data were used to convert monthly prices to December 1994 prices, and then the December 1994 cedi prices were converted to U.S. dollars using that month's exchange rate.

8   In addition to direct payments for water, those connected must pay for the connection — usually around $70 though it varies with the size of the pipe — plus a fixed fee for repair and maintenance of the water system of $1.19 per month.

9   In retrospect, it makes one wonder about the wisdom of metering all high-density customers. One Zimbabwean study showed that water consumption will be about 25-40 l/c/d in poor areas whether water is metered or not (de Kruijf, 1981, pp. 21f). The combination of introducing meters and switching to upward-stepped, "lifeline pricing," together with the growing number of people per meter in poor areas, has had the unintended effect of raising the water price levied on the poor. Of course, we must not forget that meters reduce the wasteful use of water.

10   *The Herald*, 1995a, p. 4, called for "voluntary water rationing" to reduce total consumption from the current rate of 368 megaliters per day to "less than 300". *The Herald*, 1995c, p. 6, reported that the reservoirs of Harare (and Chitungwiza) had less than 200,000 megaliters left and quoted Director of Works Tongai Mahachi as saying that we are "gambling with our lives, or at least the lives of those without boreholes." (More on boreholes shortly.) *The Herald*, 1995d, p. 4, announced that the City Council "will" impose water rationing in the next two weeks if residents do not voluntarily reduce their consumption. *The Herald*, 1995e, p. 5, declared that water rationing "is imminent as daily consumption continues to soar" and that surcharges will be imposed.

11   The estimates of the real capital stock are crude, but they are probably plausible approximations. The method by which they were derived is explained in Appendix B. Data sources: COH, Annual, various years;

COH, City, various years. (Note, in Figure 3.3, that the estimated capital stock is measured on the left-hand axis, and production on the right-hand axis; coincidentally, the numbers on the two vertical axes are the same, but the units are different.)

12 One recent government report notes that the present supply of drinking water for Harare and Chitungwiza from the Manyame River is 172 billion liters per year but that the "present unrestricted demand on the Harare water supply" was 153 billion liters per year and that this demand is "anticipated" to grow at 8% per year (MLAWD, 1993, p. 1). The report concludes that "it is therefore a matter of some urgency to identify new sources of raw water to augment the supply" (ibid., p. 2). The new sources noted by the report include some diversion of water from agriculture and much tertiary treatment and subsequent reuse of wastewater.

13 If the incremental capital-output ratio is around $1/m^3$, then the annualized cost of that must include depreciation (say .02) and real interest (say .08), making the annualized capital cost around $0.10/m^3$. For a discussion of the concept of long-run marginal cost, see Appendix C.

14 In 1994, water sales were $17.29 million, and water production was 116.83 million m^3 (COH, City, 1994). $17.29/116.83 = $0.15.

15 We use the same capital stock estimate as earlier and charge 2% depreciation and 8% interest per annum on it. Actual interest and amortization payments on borrowings are included in the "operating" expenditures, so there is at least some double-counting in our procedure.

16 There are the pumping costs but, especially since many of the City's boreholes are powered by windmill, the variable costs are extremely low.

17 The City does require a permit to drill a borehole within the City limits, but these are quite freely issued, providing the City with no more than a register of boreholes (and we could find no evidence of even this register).

18 In Gaborone (Botswana), the charge for borehole water within the City is effectively infinite since permits are rarely issued — the City sees the finite aquifer as an emergency resource to be preserved.

19 According to the 1991 Census, 93.9% of the households in greater Harare have water-flushed toilets — 92.9% in Urban Harare, 99.0% in Chitungwiza (CSO, 1994, p. 91).

20 Perhaps as many as four fifths of the houses in the low-density areas have septic tanks (World Bank, 1989c, p. 66). These septic tanks are cleaned

by the City, on request, subject to a fee of $24, which is supposed to cover cost. Since half of the City's eight suction trucks were idle on average throughout 1994, it is not obvious that maintenance costs are covered, not to mention depreciation and interest (COH, Annual, 1994, p. 132). As the number of septic tanks cleaned by the City has declined during the 1990s, two private-sector septic-tank cleaners have entered the activity.

21 The Blair toilet — like the Kumasi VIP latrine in Ghana — is essentially a basic pit latrine with a vent pipe. The cooler, fresher air enters the pit through the squat hole and the warmer, smellier air gets sucked up the vent pipe. Flies seek the light at the top of the vent pipe, are trapped by a fly screen there, and eventually die (Ziracha, 1989, p. 27).

22 There used to be a small charge for using public toilets, but it has been rescinded at all but a few, where a charge of $0.006 per use is still levied (COH, Master, 1993, p. 72). The City also maintains free public toilets — currently a total of 121 of them — around the big markets and bus stations (COH, Annual, 1994, p. 132).

23 For both systems, a yard water-tap is included in the cost. The operating cost difference largely resides in the different quantities of water needed. For Harare as a whole, there is an average of 60 liters per capita per day (l/c/d) of sewage generated, so that even in low-income areas where flushing is more controlled, something like 40 l/c/d is likely (de Kruijff, 1981, p. 30). For a household of five people, this would mean the consumption of from six to nine extra cubic meters per month (m^3/mo) of water. As argued in the preceding section, the social cost of providing this water exceeds the private cost, so water-flushed toilets generate negative as well as positive externalities. The social costs of sewerage also exceed the private sewer charges (to be discussed shortly).

24 See Appendix B for the method of estimation of the capital stock.

25 This has been especially evident in the poorest of the sewered low-income areas, partly because so much newspaper is used as toilet paper (Mafico, 1991, p. 114; Potts, 1994, pp. 212f).

26 If there are more than 200 persons per hectare (ha), pit latrines may endanger groundwater (de Kruijff, 1981, p. 30). For a discussion of other housing standards that are premature and excessively costly to the very poor, see Rakodi and Mutizwa-Mangiza, 1990, pp. 18f.

27 The trucks and crews operate on a fixed route, and they are finished whenever they have completed the day's assigned route, a common

practice around the world, but hardly an incentive to good service. Apparently they are well monitored, for there are few complaints.

28 The quoted words are those used by a City official. Actually, there are two residential areas in Harare that do not receive any door-to-door refuse pickup (Epworth and Mbare).

29 There are *four* times as many people in street-cleaning as in the entire solid-waste operation. There were also 18 vehicles attached to the street-cleaning division in 1993, up from four in 1992.

30 The expenditure data in Figure 3.6 include a 10% charge for interest and depreciation on the estimated capital stock in each division (see Appendix B for the method of estimation).

31 Service is also provided more frequently to those that wish it, at higher refuse levies. The City will even negotiate picking up from industrial establishments, at the going private haul rates, but most of that solid waste business is done by private operators or by the industrial establishments themselves.

32 The capital stock in solid waste, again calculated by the method described in Appendix B, was estimated to be over $7 million in 1992. The real cost per capita is shown in Figure 3.6.

33 Trash generation is well known to be income-inelastic. Notice in Figure 3.7, however, that the fee for twice-weekly refuse service is no longer double the fee for once-weekly service.

34 The City also owns a small dump, Pomona, that services Tafara and Mabvuku.

35 The word "apparently" is used because some City officials think that it is nearly full: "The existing landfill sites are being used up at a rate that requires urgent acquisition of new sites" (COH, Annual, 1994, p. 133); and Golden Quarry is "about 80% filled" (COH, Master, 1993, p. 72). Most, however, say that there are at least 5-10 years left at Golden Quarry.

36 In 1987, the City sampled its solid waste and found the following composition of recyclables (as a percentage of the total waste stream):

Type of Material	Amount
Paper, Cardboard	13.5%
Metal	5.4
Glass, Ceramics	4.3
Plastics	4.3
Leather, Rubber	3.2

Textiles	3.2
Wood, Bones, Straw	2.2
Total Recyclables	36.1%

The above data are reported in Tevera, 1993, Table 8.1, p. 86.

37 Tevera, 1993, which offers an excellent and detailed picture of the scavenger economy and society, says there are "700 to 1,200 pavement and dump scavengers" in Harare (p. 83).

38 The landfill supervisor's urge to control free enterprise is partly responsible for this license requirement, but it is also premised on safety considerations. Given the bulldozer's apparent obliviousness to the people in its path, it is surprising that no serious accidents have occurred in years. For health reasons, these scavengers are not supposed to live around the dump, but many do, in order to save commuting time and fares (Patel, 1988, p. 211).

39 Also collected are rubber (especially tires), various metals, whole glass containers (but not the much more abundant broken glass), and some putrescibles (which the landfill operators euphemistically call "chicken feed" but which seem more probably destined for human consumption).

40 It is amazing what is worth saving for reuse — even second-hand nails are sold in Mbare's street markets.

41 The City earns about $1,000 per month from these winning bids (Tevera, 1993, p. 87). What the City seems not to realize is that this money comes only partly from the profits of the recycling companies; it also reduces scavenger wages by giving the winning bidder monopsony power over the laborers at the landfill. This is not "just in theory": one paper recycling firm that did not win the landfill bid is paying $0.027 per kg for waste paper, but at the landfill the scavengers are being paid only $0.013 for it.

42 Tevera, 1993, p. 94, records that "the case of the scavenger who found two gold watches at the dump a couple of years ago is told repeatedly."

4 Gaborone

with B. Oupa Tsheko

Introduction

Gaborone differs from Accra and Harare in many ways. To begin with, it is almost an order of magnitude smaller — more like 100,000 than 1,000,000; and it has grown more than twice as rapidly — around 12% per annum compared to around 5%. But most importantly, it is a new city, one that has essentially been in existence for barely 30 years.

Before Independence in 1966, when Botswana was the British Protectorate of Bechuanaland, its capital was at Mafeking — in South Africa. Gaborone was constructed as the new capital of Botswana, with its site chosen mostly because of the availability of water (Serathi, 1994, p. 111). Before the completion of its large dam on the Notwane River, it "was not completely uninhabited" but consisted of a railroad station, a hotel, and about 3,800 people (Colquhoun, 1990b, p. 4.1; Feddema, 1977, p. 1).

The "Master Plan" for Gaborone assumed that it would grow to a size of 18,000 by 1990; in fact, it reached that population within five years — and continued its rapid growth, becoming today "a somewhat planned sprawl" of over 200,000 people (Letsholo, 1982, pp. 293f; Khupe, 1995, p. 3; Colquhoun, 1990b, p. 4.4).[1]

The rapid growth of Gaborone has meant that its flow of immigrants has always outpaced the City's ability to house them. By the first census (1971), there were already 7,000 squatters — nearly two fifths of the City's population. The Master Plan foresaw, and still does foresee, "a more integrated district system in which all population classes will be represented" (Feddema, 1977, p. 1; Potts, 1994, p. 217). But reality has meant a continual battle to upgrade squatter settlements into acceptable low-income housing, areas where few or no middle-income or high-income houses are found (Mosha, 1995, pp. 6ff).

Although diamonds were throughout this period transforming Botswana from "one of the world's ten poorest countries" into an "upper-middle-income" economy, the rate at which new housing could be built in Gaborone was slowed by the government's "persistent" determination to avoid urban subsidies (Molebatsi, 1995, p. 6; World Bank, 1994c, p. 163; Colclough and McCarthy, 1980, pp. 227ff; Harvey and Lewis, 1990, p. 253). This meant that those who could not afford the full cost of proper housing were unable to move out of squatter settlements or, if they did buy, were forced to pack their houses to reduce the cost per person. The 1981 census found 5.6 persons per housing plot in Gaborone; by 1989, it was 6.7 persons per plot and as much as 11.3 in some low-income housing areas (ROB, 1983, p. 2-6; Maendeleo, 1992, p. 25; Braimah, 1994, pp. 69f).

In order to recover the full cost of servicing new housing plots with infrastructure, while at the same time getting poor people into proper housing, the government has always price-discriminated in the housing market in two ways (ROB, 1973, p. 75). The first way is that, for identical plots and services, it charges different prices to different people: high "market" prices to multiple-plot developers; "full-cost" prices to high-income and middle-income families; and low "affordable" prices to low-income buyers (Mosha, 1992, p. 14; Mosha, 1995, pp. 3f).

The second way is that different levels of infrastructure are provided for different housing areas (and we will examine the details of these different levels shortly). But even in the poorest neighborhoods, "very high service standards ... have been adopted" (Mosha, 1992, p. 14). This means that an "affordable" price often covers only 10-20% of the actual cost of the infrastructure provided, requiring "massive subsidies" from the government (ibid.; Maendeleo, 1992, p. 100; Serathi, 1994, p. 115). These subsidies, in turn, reduce the number of houses that can be serviced or built with a given budget and hence exacerbate the housing shortage. Despite Gaborone's housing progress, much truth remains in the statement of more than 20 years ago that

future residential development will have to cost, on average, much less than it has in the past if the townspeople are to be able to afford it, and if heavy subsidies and large-scale squatter settlements are to be avoided. (ROB, 1973, p. 69)

70

To help overcome this dilemma between affordability and cost-coverage, the government introduced in the 1970s the Self Help Housing Areas (SHHA) to provide: 1) site-and-service plots, on which new owners would build their own dwelling, with infrastructure standards that are "watered down a bit"; and 2) "upgraded" rather than demolished squatter areas (Mosha, 1995, p. 5). Today, nearly one half of the housing plots in Gaborone are in SHHA.[2] Reassuringly, "... no significant squatter settlements have developed since the [SHHA] programme was instituted" (Mosienyane, 1995, p. 1).

The idea of SHHA has not changed much over the years. The government prepares the plots and does the roads, streetlights, storm drains, and water and sewerage mains. The new residents get a title of sorts, get a ventilated pit latrine constructed entirely at SHHA expense, are offered a small loan (to help buy building materials), and must complete a basic house within a certain time (Serathi, 1994, p. 124; Potts, 1994, p. 217). The resident pays a service levy of $4.50 per month, presumably to cover the initial infrastructure costs and the costs of the service provisions (mostly the collection of solid waste and the delivery of standpipe water).[3] However, given that the initial cost of the SHHA site-and-service infrastructure is more than $700, the service levy clearly does not currently cover all costs (Mosha, 1995, p. 11).[4]

Drinking water

In Gaborone, as in the other cities in Botswana deemed large enough and dense enough to receive unsubsidized service, the storage, treatment, and distribution of drinking water have been undertaken by the parastatal Water Utilities Corporation (WUC) since 1970 (Harvey and Lewis, 1990, p. 255). Most of Gaborone's water comes from the Notwane River, impounded in the Gaborone Reservoir, a few kilometers south of the City. Throughout the City, anyone can receive piped, treated drinking water from the WUC.

Currently, however, only about half of the City's houses get their water through a piped house connection. The difference is stark. Roughly half of Gaborone's houses are SHHA, almost all of whose residents fetch their water from public communal standpipes; and almost all of the other half of the houses have in-house or in-yard piped water.[5]

How one gets water has a lot to do with how much water one gets. One study estimated that those who walked to communal standpipes used 15 liters per capita per day (l/c/d) of water, those who had in-yard taps used 60 l/c/d, and those who had in-house pipes used 150 l/c/d (Arntzen and Veneendaal, 1986, p. 151). But casual observation suggests that the 15 l/c/d figure is far too low. Other studies have maintained that the roughly half the people who use standpipes consume barely one seventh of the domestic water (ROB, 1987, p. E.1; Arup, 1991, p. 17; GOB, 1991b, p. 4-4). But this also seems to exaggerate the differences between rich and poor.[6] Probably more accurate, and certainly more consistent with aggregate water consumption data, are the data of Table 4.1, which suggest that the poorest half of Gaborone's families consume about 60 l/c/d of water, about one third of the City's total domestic consumption.

Total water consumption in Gaborone is just under 200 l/c/d (see Figure 4.1), with consumption about equally divided between domestic consumption (including that from standpipes) and the consumption by commerce, industry, and government.[7] This per-capita water consumption of about 185 l/c/d has been quite stable for the last decade, but Figure 4.1 indicates that dramatic changes occurred in the early 1980s.

Gaborone experienced serious drought during 1984 and 1985, and this shocked the policymakers into a recognition that water was neither plentiful nor dependable in Botswana. Water consumption in Gaborone had averaged 256 l/c/d from Independence through 1983, but the supply fell to barely half that flow in 1984 and 1985. Both price and non-price policy measures were introduced in order to bring demand down to the suddenly curtailed supply.

The changes in the real water price structure since 1978 are shown in Figure 4.2, for several different levels of consumption.[8] The nominal water tariffs did not change between 1977 and 1982, even though the price level rose by 75% over this period (Harvey and Lewis, 1990, p. 256). By 1986, however, an entirely new real tariff schedule had emerged. The real price at the lowest consumption band (from zero to ten cubic meters per month) had not changed, but the degree of progressivity in the stepped tariffs at higher consumption levels had greatly increased (Gibb, 1988, p. 2-5; Arntzen, 1994, p. 107). Moreover, tariff changes since then have maintained this real price structure.[9]

Several non-price measures were also introduced during the drought: 1) irrigation and garden watering were banned; 2) the use of swimming pools and washing of vehicles were restricted; 3) the City stopped supplying water connections in new buildings; and 4) the City mounted a large public relations campaign aimed at further voluntary reductions in water use (WUC, 1984, p. 13; GOB, 1991d, p. IV-15-7; Gibb, 1988, p. 2-4; Arntzen and Veneendaal, 1986, p. 155).

As one might expect from these price and non-price policy changes, the principal impact was on the high-income households — even after the sizable real price increases, only 15% of Gaborone households spent more than 3% of their incomes on water and only 5% spent more than 5% of their incomes on water (Arup, 1991, p. 52). One study found that the "water restrictions"

Table 4.1
Gaborone water consumption by type of housing

Type of House	Percent of population (%)	Liters per capita per day (l/c/d)
All Low-Cost Housing	81	68
SHHA	62	57
Other	19	102
Medium-Cost Housing	12	127
High-Cost Housing	7	346
Average		94

Note: The average l/c/d is weighted by population.

Source: GOB, 1991b, p. 4-9.

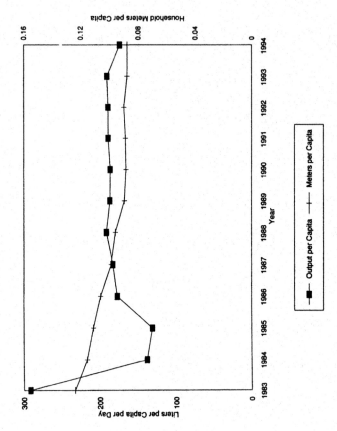

Figure 4.1. Water sales and meters (in l/c/d and meters/pop.)

74

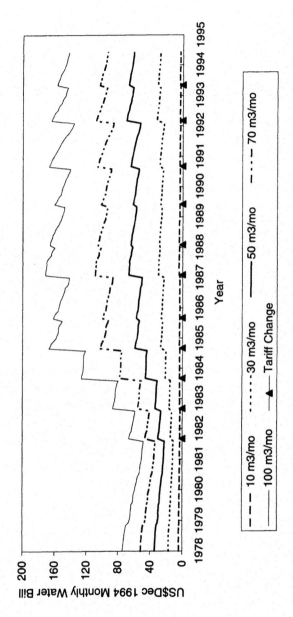

Figure 4.2. Monthly water bills, 1981-95 (in US$ Dec 94 per capita)

75

reduced water consumption by 42% in high-cost housing, by 17% in medium-cost housing, and not at all in low-cost housing (Wilkinson, 1986, p. 7).[10]

For the other half of Gaborone's residents, public communal standpipes are the water source. The WUC builds and maintains these standpipes, charging the Gaborone City Council (GCC) for the water at the lowest tariff (i.e. that for 0-10 m^3 per month). There are no attendants at these standpipes, and water there is free — free, that is, in the sense that there is no direct money price.

For SHHA housing, the rules for standpipe placement are that each standpipe will serve no more than 20 households and that no household will be further than 200 meters from a standpipe (WASH, 1986, p. 9; Colquhoun, 1990b, p. 4.11).[11] In practice, however, some houses are as much as 300 meters away, and the housing has often become more crowded since the standpipes were installed. Observation and conversation verifies that almost all standpipes are functioning, neat, and not left running.

SHHA households can become attached to the water system if they want. It costs them, initially, the connection fee and, thereafter, the meter charges.[12] Once a household becomes connected, the service levy is reduced by $0.37 per month. At this price tradeoff, only 6% of all SHHA houses have chosen to become connected (Braimah, 1994, p. 73). Indeed, in one of the poorest of these areas, Old Naledi, only 47 of the 1705 plots (less than 3%) have their own yard taps.

While the switch to in-yard water is not cheap, the fact that so few have made it is strong indication that, at low income levels, standpipe water is not terribly inferior. Although there seem to be no studies of how much SHHA households would be willing to pay for in-yard water, one survey found slightly less than half of the respondents willing to accept a "smaller plot" in return for their "own water supply" — they seemed much more willing to pay for closer standpipes and shorter standpipe queues (Mason, 1979, p. 33).

There are three efficiency problems with delivering water through communal standpipes. The most obvious one is that water is *socially* more cheaply delivered to houses by pipe rather than by bucket. Second, while SHHA residents pay for standpipe water with their time and energy, this private cost is less than the total social cost of their water. And third, rationing water by wasted time is much inferior to rationing by money price, since the latter uses no social resources.

Ultimately in Gaborone, as in all developing country cities, everyone should receive in-house piped water. The relevant question is not whether, but when. The answer is that each household should switch to piped water when it is willing to pay the marginal social cost of the switchover.[13] There are two relevant marginal social costs, that of the connection and that of the water itself. The Gaborone pricing structure reflects an understanding of the former marginal cost, but not of the latter.

The marginal connection cost, once the water mains are in place, as they are everywhere, is simply the cost of laying the pipes from the main to the yard or house and installing a meter. That is the correct price of the connection, and that is the WUC price. There are no external effects to be considered.

The marginal cost of the piped water, however, is a much more complex conception. Consider a household that switches from using standpipe water to in-house water and continues consuming the *same* quantity of water as before (an assumption that is wrong factually but useful to assist the reasoning for a while). What is the marginal social cost of this water? Zero, since there has been no change in either the water produced or in the costs of delivering it (other than the connection cost, which has been paid). The price for this water, up to the quantity that the household was consuming previously from the standpipe, should be zero.[14] In practical terms, this means that for SHHA housing, where the City is committed to providing free standpipe water, the water price should be zero at the lowest consumption-band.[15] Whether that band should be 0-5 m^3 per month or 0-10 m^3 per month depends, in principle, on what the average SHHA standpipe-using household consumes.[16]

Finally, lest it be overlooked, we should notice that, while the switch from the standpipe to piped water does not add social cost for the water, neither does it save any social cost for the water. No reduction of the service levy is called for, *if* the lowest water-consumption band carries a zero price. In practice, however, a zero price may be hard to implement, especially if it were applied as an overt subsidy to SHHA housing only. An alternative that achieves the exact same effect is to retain the present system of a basic price everywhere for water in the lowest consumption-band but to rebate the service levy by the amount of the cost (at that basic price) of the average standpipe-user's water consumption. As a guess, say that this average is exactly 5 m^3/month; this translates, at the current tariff schedule, to $1.58 per month, as compared to the present service levy rebate of $0.37 per month. The same effect as a zero price in the lowest band can be achieved by raising the service-levy rebate to piped-water users to more than four times its present level. Recall that the current service levy is $4.50, so this change would mean raising the percentage rebate from 8% to 35%.

With a connection charge equal to actual cost and a water price of zero (up to the average standpipe quantity), households would be induced to make correct social decisions about connecting. If a household's willingness to pay for piped water — i.e. the present value of the time saved and the added convenience — exceeds the connection charge, the household will want to connect, and that will be the correct social choice; and if the willingness to pay does not exceed the connection charge, the household will not want to connect, and that, too, will be the correct social choice.

Currently, Botswana is beginning to implement a change in its SHHA servicing, to require all new housing to be fully serviced with a piped, in-house, water connection (and on-site water-borne sanitation and electricity; GOB, 1991b, p. 2-24; Mosha, 1992, p. 11; Mosha, 1995, p. 16; Khupe, 1995, p. 7). The discussion above suggests that this move may be premature.[17] If the correct service-levy reduction were introduced, we guess that only a small fraction of the SHHA standpipe users would switch to in-house water. For the rest of the households, forcing them to accept such connections means one of three things: 1) these households are worse off with metered water than they would have been had unpriced standpipe water been available to them; 2) the households will be so much worse off that they will choose to pack more tightly into the existing, less expensively serviced, SHHA housing; or 3) massive government subsidies will be required. If the last route is chosen, the question arises: why does the government not also offer the same subsidy to all by paying (part or all of) the WUC connection charges throughout SHHA housing?

It is interesting to look at the budgetary implications of this suggested change — assuming it is implemented through the increased service-levy rebate — for both the WUC and the GCC. Whenever a SHHA household chooses to connect, 1) the connection charge exactly covers the WUC costs, 2) if the total water consumption is unchanged, the WUC operating costs are unaffected, and 3) the WUC collects the same revenue as before, but rather than collecting it from the GCC for standpipe water, the WUC collects it from its new water customer. For the GCC also, there is no net budgetary impact. Whenever a SHHA household chooses to connect, 1) the GCC reduces its WUC expenditures by the saving on standpipe water, and 2) its service-levy revenues are reduced by exactly the same amount.[18]

We must not forget that all the above has been premised on the assumption that the newly connected household continues to consume the same quantity of water as before. All the evidence says it will consume more. Once this happens, the WUC needs to produce more water. There are definite social costs to this additional water, and there is no justification (beyond, in SHHA housing, externality or equity) for not charging positive prices for it.

There have been two basic and long-held tenets to Botswana's urban water policy: the subsidization of standpipe water to the poor, which we have discussed, and the need for the WUC to maintain a stepped-up tariff structure that produces sufficient total revenues to cover its full costs:

In view of the scarcity of water, it is essential that this resource be treated as an economic asset. The Government is determined that the users of water should themselves meet its cost or if there must be a subsidy, that the size of the subsidy is clearly identified and made known. (ROB, 1968, p. 14)

The WUC has come very close to doing this, as Figure 4.3 shows, especially in the last five years. It has covered its operating costs in Gaborone by a wide margin in every year since 1984; but it has not quite covered full costs, once one includes an estimate of the interest and depreciation charges on the capital stock.[19]

Figure 4.3 reports cost per capita. The cost in 1994 per m^3 was $1.50. This is an estimate of the full cost of providing the current water supply in Gaborone, but it is not an estimate of the future cost of new supplies, or what is usually called the "long-run marginal cost" (LRMC).[20] The WUC has quite correctly utilized the cheaper water sources first; future expansion of water supply in Gaborone will involve ever higher-cost sources, perhaps from great distances.

An indication of the changes occurring in LRMC can be gained by looking at the movements of the WUC Gaborone capital stock and water output (sales) over the past decade (Figure 4.4). The average capital-output ratio (ACOR) over this period is 10.8 (i.e. 1994 US dollars of capital per m^3 of output), and the incremental capital-output ratio (ICOR) is 15.3 (i.e. 1994 US dollars of new capital per m^3 of new output).

Allowing 8% interest and 2% depreciation, this means that the capital costs of providing water in Gaborone are already in the range, $1.08-1.53. With operating costs at their present level of $0.47, this suggests that the LRMC has already risen into the range, $1.55-2.00. Others' estimates of the LRMC are even higher: $2.59 per m^3 (Arntzen, 1994, p. 108); and $2.42 per m^3 (GOB, 1991d, p. IV-13-3).[21]

All these estimates of LRMC are far above the average price that has been charged for water in Gaborone in the past decade. Over 1986-94, the average price has ranged over $1.21-1.40, and it was $1.24 in 1994. Others have noted that, to set average price equal to LRMC, dramatic increases in water tariffs would be needed:

The exceptional volume of capital expenditure required in the urban water sector over the next decade would, if financed under conventional loan

79

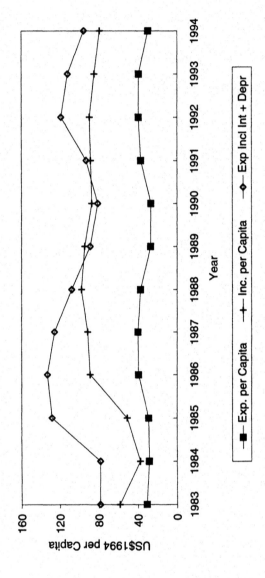

Figure 4.3. Water income, expenditure (in US$ 1994 per capita)

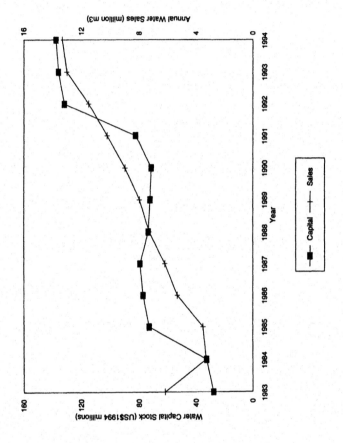

Figure 4.4. Water capital stock, sales (in US$ 1994 millions and million m3)

arrangements, lead to insupportable debt service requirements under any realistic tariff policy. (GOB, 1991a, p. 9-11)[22]

Whatever the exact figures, it is clear that future increases in water demand are either going to require big increases in (average) water tariffs or going to require big government subsidies.

The obvious question at this point is: are there any easy non-conventional ways to increase water supplies or to reduce water demands? Let us look at a few that have been suggested:

1   Increased WUC efficiency. There is general agreement that there is little scope for expanding water supplies through more efficient operation of the WUC. Unaccounted-for-water (UFW), a common measure of the degree of inefficiency in water distribution, is in the range of 6-20%, very low for a developing country city (Wilkinson, 1986, p. 9; ROB, 1987, p. 8.7; GOB, 1991a, p. 4-8; GOB, 1991d, p. IV-15-6). And the bill collection rate, another common measure of efficiency, is extremely high. When bills are not paid, the WUC quickly disconnects — one tenth of the City's meters were disconnected at some time in 1994. There is a $15 charge for reconnection, and two thirds of those disconnected meters were reconnected before the year was out. The WUC avoids the problem, common in other developing countries, of collecting from government agencies; it sends their bills directly to the Ministry of Finance, which pays the total government bill with one check and then debits the various agencies' budgets.

2   Increased groundwater use. While Gaborone is blessed with surface water — indeed, it was built there for that blessing — there is little groundwater and its rate of recharge is so low that using it would represent once-and-for-all mining.[23] The WUC thinks of Gaborone's little groundwater as an emergency resource, to be saved until desperately needed. The government implicitly agrees, having rejected all applications to drill boreholes within Gaborone for at least the last ten years. Effectively, and we think sensibly, the government charges an infinite price for borehole water in the City.

3   Increased rainwater use. There have been many mentions but few studies of this possibility.[24] One careful study concludes that it would cost $35-80 in capital costs per cubic meter of rainwater stored (Gould, 1984, p. 25).[25] At the current price of water in the lowest consumption-band, this would

mean a private rate of return of between 1/2% and 1% per annum. At an opportunity cost of the LRMC, say $2.50, the social rate of return rises to 3-7%, better but still not exciting.[26]

4 Reduced demand through education and/or prohibition. Although it is difficult to separate out the impacts of the different 1984-85 drought policies, public urgings and outright bans undoubtedly played a part. Unfortunately, all the easy reductions have occurred — the lawns that burned away then were mostly not replaced, and swimming pools are few in Gaborone. Indeed, a major reason why water demand did not go back up to its previous level in the late-1980s was that permanent reductions in inessential water usage were made by high-income consumers (WUC, 1985, p. 13).

5 Reduced government demand. Government use of water in Gaborone has been rising at 18.9% per annum over the past ten years, while non-government use has been rising at only 13.4% (see Figure 4.5). There are at present few incentives for government agencies to economize on their water use, and it is not hard to think of ways to provide such incentive.

6 Reduced zero-price water distribution (for in-house piped consumers). Many Gaborone houses are provided to employees as part of their remuneration, and often the employer pays the water bill as well. This means effectively a zero price for the user, and the consumption reacts to it; those who do not pay for their water consume nearly six times as much as otherwise identical consumers who do pay (Arup, 1991, p. 55). A major reason for the low Gaborone estimates of the price elasticity of water demand is that many people who make the consumption decisions do not actually pay the price (GOB, 1991b, p. 4-31). The government or the WUC should definitely explore ways to reduce this practice — at the very least, the water consumption could be included in household income for tax purposes.

Human waste

When Gaborone was established in the mid-1960s, practically all of it was sewered by a system of gravity sewers, with treatment facilities in a series of waste stabilization ponds (Commonwealth, 1965, p. 126; ROB, 1983, p. 2-4; WASH, 1986, p. 2; Colquhoun, 1990b, p. 4.12). While almost everyone in

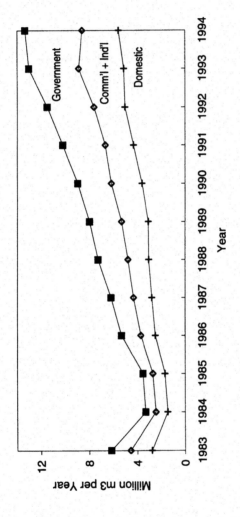

Figure 4.5. Water allocation (in million m^3 per annum)

84

Gaborone can be attached to this sewer system, not all in fact are, by any means. With sanitation, like water, there are two Gaborones, the SHHA housing and the non-SHHA housing.

The non-SHHA plots, with roughly half of the City's population, have water-borne flush toilets attached to the sewer system; the SHHA plots, with the other half of the population, have pit latrines in their yards.[27] Actually, the dichotomy is not quite that clear-cut. Many of the non-SHHA houses have preferred to remain with septic tanks; and a very few of the SHHA houses have water-flushed toilets attached to septic tanks rather than the sewer — in Old Naledi, for example, five of the 1705 plots have flush toilets and septic tanks (Khupe, 1995, pp. 14f).[28]

The core of every SHHA plot is a ventilated improved pit (VIP) latrine. Typically, it has a block-lined pit two meters deep, with a center wall to divide it into two parts, a cover slab with two openings for the seat, and a ventilation pipe. In principle, the family uses one pit until it is full, then switches to the other, letting the first "digest" for a year; this reduces both the volume of sludge and the health hazards to the cleaners (Gibb, 1989, p. 5-4). Pit latrine emptying is done by the GCC, free and on demand (Khupe, 1995, p. 15).

To empty pit latrines and septic tanks, the GCC maintains a fleet of four tankers, but half of these vehicles are typically under repair, so that there may be a long wait for service — usually a few weeks, sometimes several months (Gibb, 1989, p. 5-1; Khupe, 1995, p. 15; Braimah, 1994, p. 117).[29] The contents are emptied into the sewer system.

A sanitation system consisting of pit latrines can be a source of serious problems in crowded areas, whether the crowding is in the high number of latrines or in the high number of people using each latrine. First, heavy and careless use burdens the sludge removal system:

> The expected sludge removal frequency of a pit latrine under normal use is once in 2-5 years. However it was found that in Gaborone some pits have to be emptied once a week and many of them once in 12 months. (Gibb, 1989, p. 5-1)

Some pits are hard to empty because they have also been used for solid waste disposal. Second, the density of the housing makes it difficult, and sometimes impossible, to get the vacuum-tanker close enough to drain the pit. And finally, leachate can pollute the groundwater, although this does not yet seem to be an important worry for Gaborone.

One way to avoid these problems is to convert the houses with pit latrines to water-flushed toilets on the sewer system. Botswana is moving in that direction. The Accelerated Land Servicing Programme (ALSP), which applies to all cities, requires that all new SHHA plots receive on-site, water-borne sanitation (GOB, 1991b, p. 2-24).[30] This is certainly the ultimate goal — again, the question is not whether, but when. Three problems are faced by moving too fast:

1. Water-borne sewage is expensive. This well-known fact can be seen in many ways. The capital cost of sewers is seven times that of pit latrines (GOB, 1991c, pp. 2-5f); the capital cost of installing sewers per SHHA plot is $964 (Olsen, 1989, p. 18); the present value of all costs of operating a water-flushed toilet is nearly eight times that of a pit latrine — $5,885 versus $750 (WASH, 1986, pp. 20f; Gibb, 1989, p. 5-3); and the total cost of SHHA infrastructure is reduced by 18% if pit latrines and standpipes are used instead of in-house water and sewerage (Olsen, 1989, p. 68).

2. Endemic sewer blockage could result. In developed countries, about 80% of water sales ends up discharged into the municipal sewer, but in Gaborone the return flow is more like 50-65% (ROB, 1983, p. 3-22; HIFAB, 1987, p. 29; GOB, 1991c, p. 2-3).[31] Already, with only the higher-income half of the City on sewers, people are finding ways to economize on water — e.g. by flushing only after several uses. Guidelines for sewer operations usually call for 50 l/c/d to keep waste flowing; for a family of five, this is 7.5 cubic meters per month of water, just to keep the sewer flowing.[32] However, 7.5 m^3 is more than most families in Gaborone consume per month (GOB, 1991c, p. 2-2). If poor people were added in great numbers, the concomitant reduced sewer flow "could well upset the entire operation of the local sewer system" (GOB, 1991c, p. 2-3).

3. Water-borne sewage means increased water consumption. Even if poor families economize to the utmost in their use of water, there is no way that they can avoid increasing their consumption. The "typical" water usage by type of toilet has been estimated to be 35 l/c/d for those with a pit latrine and 75 l/c/d for those with a conventional water-flushed toilet (GOB, 1991c, pp. 2-5f). Back-of-the-envelope calculation tells us that, if the 50,000 SHHA residents all switched to water-borne toilets and consumed 20 more l/c/d each, Gaborone would need to supply 1,000 more cubic meters of water per day — that would mean a 7% increase in the City's current rate of domestic consumption. To carry this illustration

a step further, if all that water found its way to the sewers, the treatment system would have to process more than 10% more sewage than it does now.[33]

Forcing households into sanitation systems that they cannot yet afford means either that these households will be worse off with water-flushed toilets than they would have been with pit latrines (with both lower capital cost and concomitant lower water usage) or that massive government subsidies will be required. As one report put it:

> The record of conventional sewerage systems for low income/high density communities in developing countries is not encouraging. (GOB, 1991c, pp. 2-5f)

In short, there seems to be no greater willingness by really poor people to pay for in-house toilets than for in-house water (Mason, 1979, p. 33).

As with the switch from standpipe to piped water, the switch from pit latrine to water-flushed toilet should basically be made when the household becomes willing to pay the social costs of the switch. Government policy may deem a subsidy appropriate to encourage more rapid switchover, but one should in any case know what the cost is — if only so that one knows the extent of the subsidy being granted.

There are three aspects to the social cost of a switch from pit latrine to waterborne sewage:

1   There are the hardware costs of pots and pipes (from house to sewer). These are completely borne by the household and hence are fully considered in the private decision to switch. In many countries, where capital markets are believed to discriminate against small borrowers, the government offers low-interest loans — and in Botswana, there is some subsidized "stretching out" of the payments through the Botswana Building Society.

2   Water consumption will almost certainly increase, and many of the switching low-income (or even middle-income) families will be buying water at a price below LRMC. This means that the private decision does not fully consider the water cost imposed on society by the switch.

3   The net operating cost to society of treating the waste must also be considered. There is a marginal cost to maintaining the sewer, none of

87

which is directly paid by the household. And there is a reduction in marginal cost of operating the pit latrine cleaning service, none of which is paid by the SHHA household.[34]

The sewer flow in Gaborone is currently 18-25 thousand cubic meters per day. This waste is deposited in stabilization ponds, covering 52 hectares, where treatment occurs through natural processes, with no machinery or energy (except sun) input. The result is a reasonably high standard of treatment, although the effluent is not suitable for drinking. Some of the effluent is used for irrigation, while some is further treated and discharged back into the Notwane River (ROB, 1993, p. 1; Masundire, 1994, p. 153; Khupe, 1995, p. 11).

These treatment facilities have been called "below acceptable standards ... [with a] high risk of pollution" for the Notwane River "in the near future" (ROB, 1993, pp. 1f). As a result, and at great expense, a new treatment plant will shortly replace the stabilization ponds, supposedly producing an effluent that can be reused as an additional water source for the City (Khupe, 1995, p. 12).

The switch to sophisticated sewage treatment may have been premature (Gibb, 1983, p. 4-1; Khupe, 1995, p. 12). It is very costly, especially where land is plentiful, temperatures high, and hence stabilization ponds particularly cheap and effective (Gibb, 1983, p. 3-36). The Botswana National Water Master Plan considered a 30 thousand m^3/day treatment plant, and advanced treatment options all ran two to three times as expensive as waste stabilization ponds (GOB, 1991c, p. 4-9). But one should take these comments with a grain of salt, for Gaborone's real expenditure on sewerage has been declining over the past decade from around \$4 per capita to around \$2 per capita (see Figure 4.6).[35]

Solid waste

The Health Department of the City of Gaborone assumes responsibility for picking up basic residential and commercial solid waste, leaving construction rubble, garden waste, etc. to the private sector.[36] Typical among cities in developing countries, Gaborone generates about 0.5-0.6 kg/c/d of basic waste, two thirds residential, one third commercial (Tevera, 1994, p. 22; Khupe, 1995, p. 16; Dohrmann et al., 1991, p. 3-2; Dijeng, 1994, p. 1).[37] Also typically, it is dense waste, about 250 kg/m^3 (COG, 1993, Figure 3-1).[38]

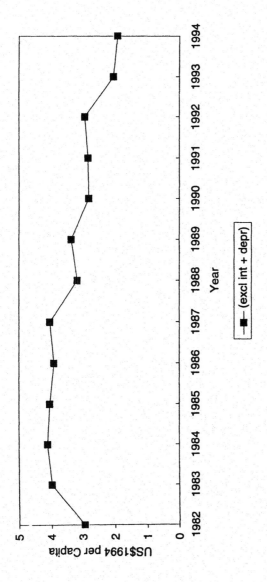

Figure 4.6. Sewerage expenditure (in US$ 1994 per capita)

Collection is free — in the sense that there is no direct money charge — and is made at least twice weekly in the residential areas and as often as necessary in the commercial areas.

Until 1994, the entire City was serviced by the GCC. But last year, at the urging of the U.S. Agency for International Development (USAID) and with its financial assistance, the basic collection service was privatized for about half of the City (Gaborone South). It is too early for any formal comparison of the efficiency of the two operations, but a look at the differences in the techniques of operation is instructive.

The public-sector operation utilizes a variety of vehicles and techniques, ranging from tractor-pulled trailers carrying the forty-odd "skips" (i.e. large bins) from the commercial areas, to round-bodied and square-bodied trucks collecting from the low-income areas, to small modern compactor-trucks servicing the high-income areas, to open trucks accumulating the black plastic bags filled by street-cleaners, litter-pickers, and park-workers (Khupe, 1995, pp. 17f).[39] In recent years, the GCC has maintained some 20-40 vehicles for solid waste collection, though about one third of the vehicles are not operating at any given time, which means that the actual collection is sometimes "infrequent" (Braimah, 1994, p. 119; Dohrmann et al., 1991, pp. 3-5, 3-8).

The collection is from skips in the commercial areas, but there is a shortage of these and hence many get overfilled and surrounded by litter. The collection is from barrels at the individual plots in residential areas. In SHHA and BHC housing, the initial containers are provided; elsewhere, the owners must provide their own containers. The result is that there is theft of containers, some are bottomless, and most are old and lidless. They tip over easily, and the scavenging dogs seem talented at encouraging such tipovers.

These litter problems are compounded by the fact that Botswana's spectacular growth over the last two decades has begun to move it into the "throwaway" stage of development, between the low-GDP-per-capita paucity of trash-generation and the high-GDP-per-capita sophistication of trash-treatment. In Gaborone, the change is most clearly seen in beverage containers and plastic shopping bags. The beverage containers are no longer glass, no longer with deposits or high reuse-value, but not yet high-value aluminum, not yet much recycled — and hence much littered. And most shops now routinely provide plastic bags for customers, who react by no longer bringing their own reusable bags.

The "throwaway" stage is clearly one of more than marginal change — one careful observer believes that the total solid waste generation has more than doubled in Gaborone since the late 1980s (Dijeng, 1994, p. 1). There are frequent professional complaints about: the growth of "indiscriminate waste disposal ... [and] open spaces left for recreational purposes ... [being] used as dumping sites" (Molebatsi, 1995, p. 19]; the fact that the planned open spaces in the City "have ended up as dumping grounds for rubbish" (Mosha, 1995, p. 13); and the "unsightly situation of our roadsides and highways .. [which shows there is] neither care nor sensitivity to the problem of littering" (Segosebe and van der Post, 1991, p. 7). "Campaigns are mounted time and again to clean the city" (Mosha, 1995, p. 22); but "law enforcement to deal with this [litter] problem is very lax" (Khupe, 1995, p. 23). This is the same country where the appearance of one littered Coca-Cola bottle once created a major motion picture.

The private-sector operation, Daisy Loo, also uses different pickup techniques in different areas:[40]

1   In its commercial areas, Daisy Loo, like the GCC, places skips, and its two big compactor-trucks pick these up very early in the morning — partly to avoid traffic congestion later and partly to prevent complaints about early-morning noise in the residential areas.[41]

2   These same trucks do twice-weekly, curbside service through the residential areas, picking up all neatly contained solid waste (either bagged or barreled).

3   In Old Naledi, where the roads are narrow and many houses are off-street, the big compactors experience difficulty in operation. So throughout the entire area, Daisy Loo has sub-contracted the initial solid waste pickup to the 18 Ward Development Committees (WDC). Each WDC hires a handcart operator to go door-to-door and carry the trash to one of the two skips that Daisy Loo maintains in each ward. The big compactors simply collect from these skips, roughly twice a week, but more often if they get full.[42]

When we compare the public and private operations, what is most striking is the absolutely opposite dynamics with respect to factor-intensity. The public operation has been attempting to substitute capital-intensive compactors for traditional trucks that are less expensive, less prone to breakdown, and less complex to repair. Meanwhile, the private operation, having been forced to start with huge, state-of-the-art compactors, has been attempting to substitute labor-intensive ancillary activities for their use.[43]

Throughout the residential areas of Gaborone, (at least) twice-weekly service is standard. City officials maintain that this is essential for health

91

reasons in a hot climate, but it does nearly double the cost of the entire operation.[44] In most of the world, even in much richer (and also hot) countries, once-weekly pickup seems generally acceptable. From casual observation, one wonders whether spending more picking up the trash that does not reach the barrels would not be a better use of the budget available.[45] Alternatively, reducing the budget for solid waste collection and reducing the service levy in SHHA might be considered.

All the solid waste collected, publicly and privately, goes directly to the landfill, sited at the southeast edge of the City. It is surprisingly close to the Notwane River (though below the reservoir), a placement justified on a variety of grounds: that most of the solid waste is inert matter; that the relatively low rainfall produces little leaching action; that the adjacent sewage ponds would pollute the underlying aquifer anyway; and/or that the aquifer is already "of poor quality" (ROB, 1991a, p. 24).

The old landfill, closed in 1993, was truly a "dump" — there was essentially no effort to build cells or to provide cover, and flies, odor, rodents, litter, fires and the disposal of hazardous wastes were rife (Dohrmann et al., 1991, pp. 3-15ff; Tevera, 1994, pp. 21ff). Its closing was welcomed, though apparently not done in a very careful manner (Somarelang, 1995, p. 3).

The new landfill is more appropriately (i.e. more carefully but not excessively) managed. There are several parts to it, with tires, construction debris (and other inert matter), and brush and tree cuttings being segregated into less safeguarded portions.[46] The main body of the landfill is covered daily, fires are fewer, and leachate is collected (Dijeng, 1994, p. 1; Khupe, 1995, p. 18; Molebatsi, 1995, p. 19).

Because of the categories in the GCC budget, it is not easy to estimate the overall costs of the City's solid waste collection and disposal. We have tried to do that in Figure 4.7, which extends a previous study (Dohrmann et al., 1991, p 3-24).[47] The real solid waste expenditure per capita seems to have been falling slowly over the last decade and is now around $5 per capita per year — which means about $20-30 per ton.[48]

Scavenging for recyclables has long been practised in Gaborone, but it has become a more organized activity since the new landfill opened. Before 1993, there were from 150 to 600 "uncontrolled" scavengers working the unfenced dump each day, living at its perimeter; now, there are less than 100, and they are no longer permitted to live there (Dohrmann et al., 1991, p. E-2; Tevera, 1994, pp. 25ff).[49] Before 1993, the scavengers were seeking little else than scrap metal and food for themselves and families; now, they collect cans, cardboard, paper, and some plastic as well (Dohrmann et al., 1991, p. 3-17).[50]

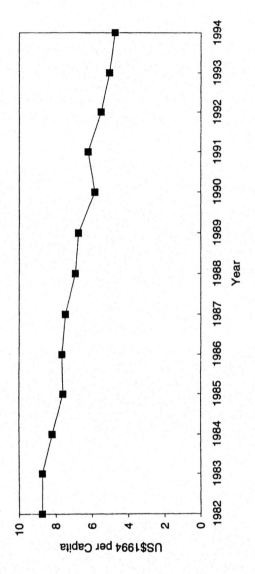

Figure 4.7. Solid waste expenditure (in US$ 1994 per capita)

Cardboard, paper, and plastic bags are recovered from the landfill by Waste Paper Recovery, which sends regular trucks to pick these materials up. They are sorted, cleaned, and baled at their Gaborone materials recovery facility (MRF). The paper is sent to Harare, the plastic to Johannesburg. On the average, the firm collects about ten tons per day of paper and five tons per month of plastic. These are, by the very fact of Waste Paper's operation, commercially viable quantities, but they do amount to only 0.05 kg/c/d of paper and "trace" amounts of plastic.

What is unique about Waste Paper's operation is that it pays its scavengers *by the hour*. Nowhere else in the developing world, to our knowledge, is this done. The firm has a half dozen regular employees at the landfill (and others stationed around the City where large quantities of paper appear), all of whom are paid at least the minimum wage and work regular hours.

The explanation for the hourly wage is interesting. Years ago, the firm paid the traditional piece rates to scavengers, but they found that the collection rate was too erratic — on some days the scavengers worked hard and on other days not at all. This meant that the salaried workers at the MRF ended up with overtime some days and no work at all on others. It also meant that, on the days scavengers were not working, much of the potentially recyclable paper was burned away in the frequent dump fires. The switch to regular wages and hours solved both problems, and the anticipated new problem of monitoring work effort was easily overcome.

The other major recyclable in Gaborone, steel cans for food and beverages, is collected from the landfill — and elsewhere in the City — by Metal Box (Botswana). This firm sends trucks anywhere in the City, including the landfill, where large quantities of cans appear; it also accepts cans delivered to its MRF. It pays $0.07 per kg for steel cans and $0.37 per kg for aluminum cans if they are delivered to the MRF — and about half of those prices if Metal Box has to go to collect them (Somarelang, 1995, p. 7).[51] These prices are nearly twice (in real terms) the prices of a few years ago (COG, 1993, p. 6-1). The firm collects about ten tons per day (i.e. 0.05 kg/c/d), which is about 10-20% of the cans sold in Gaborone (COG, 1993, p. 6-2). These are compacted, baled, and sent to Pretoria.

While there is not nearly so much recycling in Gaborone as in most developing country cities, one should recognize that Botswana is under two severe handicaps in this respect. Market-oriented recycling is seriously disadvantaged by sparse population and by long distances. Profitable recycling requires that large quantities can be collected within a small area, and it also requires that processing plants of minimum efficient scale (MES) can be located near the recyclables (since most have high weight-to-value ratios).[52]

The question often arises in Gaborone: should the government actually encourage recycling beyond what the market has provided? The policy most often discussed is the introduction of a City owned and operated MRF at the landfill. This has been recommended by USAID and GCC studies made at the same time as the privatization of part of the solid waste collection (Dohrmann et al., 1991; COG, 1993). The MRF is supported in these studies partly from an urge to gain more control over the recycling activities, partly from a concern for the unhealthiness of the scavenger's activities, and partly from economic analysis (Dohrmann et al., 1991, pp. 4-8ff, E-2).

The Dohrmann et al. economic analysis is, however, seriously flawed. It correctly points out that a full-scale, modern MRF could recycle much more material than is currently achieved.[53] But it fails to calculate accurately the economic cost of such recovery. Although they compare several recycling/composting/incineration alternatives, the critical choice is between a straight landfill with no ancillary (public) activity ("alternative 1") and a landfill with a 15 tons-per-day (TPD) waste-sorting and 27 TPD composting MRF ("alternative 3"). They conclude that alternative (3) has "the lowest comparative cost," although their own numbers show alternative (1) as having the lowest cost (ibid., p. 4-16).[54]

But the Dohrmann et al. computations are not the right ones. The correct formula for the annualized cost is

$$\text{annualized cost} = \text{(net recurrent cost)}$$
$$+ .08 * \text{(landfill construction cost)}/(1-[1.08]^{-N})$$
$$+ .08 * \text{(all other capital cost)}/(1-[1.08]^{-25}),$$

where net (of any recycling revenue) recurrent cost, landfill construction cost, and all other capital cost are their estimates, N is the life of the landfill they calculate under the alternative being considered, 0.08 is their assumed discount rate, 25 years is their assumed life for all equipment used (including vehicles) — and their assumption is accepted that the rate of solid waste generation will not grow over time.[55] With this corrected formula, alternative (1) is even lower in cost relative to alternative (3) — 26% lower, whereas the earlier calculations found it to be 8% lower.[56]

A later report admits that the MRF costs "appear high, relative to landfill costs" but points out that recycling revenues "are projected" to rise faster than operating costs in the future. This, of course, is not an argument for building a MRF now but for planning to build it at some unknown time in the future — MRF construction is not a now-or-never decision (COG, 1993, p. 6-8).

In short, MRFs are expensive propositions, and recycling revenue does not come close to covering the expense — as communities all over the United States have been recently discovering. They make economic sense when, and only when, the cost of landfilling has become extremely high, either because the land itself is scarce or the groundwater beneath it is seriously at risk. In Botswana, land is plentiful and groundwater scarce. The MRF is an operation whose time has not yet come.

The very factors that make Botswana infertile ground for recycling — sparse population and long distances — have brought about an early switch from returnable, reusable, deposit-bearing beverage containers to "one-ways."[57] Most beer is now sold in such containers, which helps to account for the heavy litter on roadsides and in other public spaces.

Various policies have been suggested to counter this trend to one-way containers, or at least to counter the trend to the concomitant increased litter. Environmental groups have been encouraging breweries to return to returnable, reusable bottles, but probably inevitably without success (Khupe, 1995, p. 18). Other groups have been pushing for a mandatory deposit on beverage containers, also so far without any success. Aside from the absurdly low deposit they have suggested ($0.004 per can or bottle), mandatory deposits are an expensive way to control litter, probably not yet one for which the average Botswana citizen is willing to pay.[58]

Summary

Gaborone is a small, planned capital city, although it has grown far more rapidly than its planners envisioned. This continual rapid growth has meant that the flow of migrants has always outpaced the City's ability to house them. The City has attempted to meet the needs of the poor migrants, within budget constraints, through cross-subsidization and the provision of different service levels to different consumers.

About half of Gaborone's residents have piped in-house or in-yard water connections; the other half relies on a system of regularly spaced public standpipes. Those who utilize standpipes consume much less water, about 60-70 liters per capita per day (l/c/d) as compared with 130-350 l/c/d in medium-cost and high-cost housing.

Since the drought of the mid-1980s, Gaborone has dramatically raised its real water prices, and consumption, especially in high-cost housing, has been much reduced. But new water supplies in Gaborone will indeed be very expensive, and even higher real prices may be needed to ration it efficiently. Standpipes are many in the poor areas, and the water from them is free. The move to require poor households to accept in-house water connections may be premature, given the costs and their willingness to pay.

Gaborone is fully sewered, but roughly half of its households have chosen not to attach to it, instead maintaining pit latrines in their yards. Pit latrine emptying is done free and on demand, but there is increasing interest in forcing sewer attachment because of the high density in the poor residential areas. This switch may be premature.

Solid waste is picked up for free and at least twice a week throughout Gaborone. This collection is done partly by the City and partly by a private contractor, and the two use very different techniques of collection. Litter is becoming a serious problem, and the recent trend to one-way beverage containers is worrisome. The City landfill has recently been improved, and scavenging there for recyclables has become more formal.

Notes

1   The population was 140,000 in the most recent census (1991).
2   Another one third are built, owned, and rented by the Botswana Housing Corporation (BHC). Upscale of SHHA plots, the BHC houses are provided with in-yard or in-house water connections and waterborne sewerage. The BHC is also constrained to cover its costs, so its rents are higher (Maendeleo, 1992, p. 23; Braimah, 1994, p. 69; Mosha, 1995, p. 14).
3   The service levy is to the SHHA plot as the property tax ("rates") is to all other housing. The service levy is lower in real terms now than it used to be — e.g. it was $6.60 in 1981 (van Nostrand, 1982, p. 58) — though the City now provides more services. Some of the services delivered to SHHA plots are charged for, but usually at much less than cost — e.g. for emptying a pit latrine, SHHA residents pay $2, while others pay $11.
4   Some three fourths of the SHHA residents are delinquent in the payments of the service levy, revenues from which typically contribute less than 3% of the City's budget. Property taxes contribute about two fifths, with the rest coming as a grant from the central government (Mosha, 1995, pp. 22ff). Default on the building material loans is also "widespread" (Moyo et al., 1993, p. 39).

5   Fewer than one tenth of the SHHA houses have in-house or in-yard piped water. On the other hand, almost all others do — for example, even in BHC housing, 91% of the plots have piped water (Braimah, 1994, p. 73; GOB, 1991b, p. 2-24).

6   If it were really true that households in the upper half of the income distribution consumed six times as much water per capita as the lower half, it would mean either that the income elasticity of demand for household water was greater than one or that the per-capita incomes of the richer half are more than six times (on average) the per-capita incomes of the poorer half. In Gaborone, as almost everywhere, the income elasticity of demand for household water is well below unity (Arup, 1991, p. 51; GOB, 1991d, p. IV-15-6). And income inequality in Gaborone, while high, is surely not that high.

7   Figure 4.1 also shows the number of household meters per capita, which fell in the late-1980s but has remained stable in the 1990s. The fall in the 1980s occurred partly as the result of the rising importance of SHHA housing (which added population without adding demand for meters) and partly as the result of increased crowding in non-SHHA housing (which increased the number of people per meter).

8   The ticks for the years in Figure 4.2 are shown at the mid-year, i.e. 1 July. For this graph, monthly CPI data were used to convert monthly prices to December 1994 prices, and then the December 1994 cedi prices were converted to U.S. dollars using that month's exchange rate. The water tariff schedule is the same for residential and commercial users. Tariffs are set by the WUC with the approval of the Ministry of Mineral Resources and Water Affairs. Almost from its inception, the WUC has maintained a stepped water tariff schedule, on the grounds that "equity and public health considerations necessitate some subsidy to the lowest step in the tariff structure" (ROB, 1973, p. 156; ROB, 1977, p. 175). Also, there has always been a minimum monthly charge, not shown in Figure 4.2 because even the charge for 10 m^3 per month has always exceeded the minimum charge.

9   There has been a slight change in the real price structure over the past decade — the real water bill of those consuming 30-50 m^3 per month has increased somewhat. This change has occurred principally as a result of the WUC's partial acceptance of a recommendation that "a reasonable estimate for household basic needs is 10 m^3/month" and the subsequent reduction of the base-price consumption band from 0-15 m^3 per month (where it had been over 1987-1992) to 0-10 m^3 per month. Actually, the

original recommendation was to reduce the basic band to 0-5 m^3, which would allow a family of five to get 35 l/c/d, "sufficient to meet all basic needs" (GOB, 1991d, p. IV-13-6; Arup, 1991, p. 37).

10 Since overall domestic consumption went down by more than 40%, these estimates should be used only as indicators of the relative impacts. There are few estimates of price elasticity of demand for water in Botswana, but they run around (minus) 0.5-0.7 (GOB, 1991d, p. IV-15-5; Arup, 1991, pp. 55, 63). The methods used by Arup, 1991, to get the 0.7 estimate are particularly ingenious, utilizing two natural experiments: 1) water prices differ between otherwise identical consumers in Botswana, in that different cities have different price structures; and 2) water prices differ between otherwise identical consumers in Gaborone, in that some consumers have all or some of their water bills paid by their employers.

11 From the late-1970s to the mid-1980s, the rule was one standpipe for every 15 households (van Nostrand, 1982, p. 54). Before 1977, however, standpipes were very few in some poor areas — only four for the thousands of households in Old Naledi, and some people lived two kilometers from the nearest standpipe (ibid., p. 29).

12 The amount of the connection fee depends basically on the distance to the water main since it is set so as to cover the cost of laying the pipes to the house and installing the meter. Connected households are also responsible for any repairs needed between the meter and the house — the meter is usually placed at the plot boundary (WUC, 1979, pp. 8f; Khupe, 1995, p. 7).

13 We ignore for the moment the positive externalities provided by greater water consumption. Consideration of these externalities calls for government policy to speed up the switch to piped water.

14 One government report recommended that the price of water in the basic band be set equal to short-run marginal cost (SRMC) and then calculated that to be $0.82 per m^3 (GOB, 1991d, p. IV-13-6). The idea of using SRMC is sound, but the report forgets that some of this basic water costs less because it reduces the need to provide standpipe water.

15 The thought of pricing standpipe water has occurred to policymakers in Gaborone (GOB, 1991d, p. IV-13-8). Long practice and equity considerations are, however, against it. Furthermore, given the number of standpipes in Gaborone and the fact that an operator would be needed at each, it almost certainly would not be cost-effective in practice.

16 We were unable to uncover this average SHHA water consumption figure.

17 It has been estimated that in-house water and sewerage raises the cost of SHHA plot infrastructure by 18% (Olsen, 1989, p. 68).

18 In fact, given that service levies are often not fully paid up, while the WUC payments are almost always made, the GCC would come out ahead on the switchover.

19 The estimates of the capital stock are made by the methods described in Appendix B. Readers should be reminded that the estimate is subject to a great deal of possible error and hence that Figure 4.3 is only a very rough indication of whether the WUC is covering full costs. But Figure 4.3 does follow recent WUC reports of its financial success: in 1991, "more than satisfactory" (WUC, 1991, p. 1); in 1992, "satisfactory" (WUC, 1992, p. 1); and in 1993, inadequate because of "the delay in approval" of tariff increases (WUC, 1993, p. 1). One final observation: apparently, the WUC goal is to cover the amortization of its external borrowings *and* to cover the interest and depreciation on its total capital, including that financed externally, but this would double-count much of the capital cost (WUC, 1994, p. 4).

20 See Appendix C.

21 This last estimate explicitly uses an 8% interest charge; the LRMC falls to $1.95/m^3 at 6% interest and rises to $2.98/m^3 at 10% interest (GOB, 1991d, p. IV-13-3).

22 This report implicitly estimates the LRMC at $2.11/m^3 (i.e. 70% above the current average price of $1.24/m^3).

23 For the rest of Botswana, on the other hand, two thirds of the water supply is from boreholes (Arntzen, 1994, p. 102).

24 Unlike most of Botswana, Gaborone is not a low rainfall area — it averages 550 millimeters per year (Gould, 1984, p. 3).

25 This is the "biggest cost" and other costs are not quantified (ibid.).

26 This same author later claimed that construction of rainwater tanks on large government buildings in Gaborone could yield water at a cost of $0.66/m3, which is "competitive" with WUC rates and "well below the long-run marginal cost" (Gould, 1994, p. 67). But no details are given.

27 The exact numbers are hard to estimate because so many different units of measurement can be used. Of the 60,000 *plots* in Gaborone (residential and business), 50,000 are attached to the sewer (Khupe, 1995, p. 10). Pit latrines are used by "over 8,000 *families*" (Gibb, 1989, p. 5-1). In 1986, 50,000 of Gaborone's 96,000 *people* used toilets that were attached to the sewer system (HIFAB, 1987, p. 7). The 1991 Census indicated that 34% of Gaborone *households* have flush toilets (Braimah, 1994, p. 73).

Finally, in 1991, there were 8,261 pit *latrines* in Gaborone (ROB, 1991a, Table 3.8). (We have added the italics in the above sentences.)

28 In 1991, there were a total of 3,507 septic tanks in Gaborone (ROB, 1991a, Table 3.8).

29 There is a charge of $11 for emptying septic tanks. In recent years, the GCC has emptied 220 pit latrines and cleaned 130 septic tanks per month on average.

30 Nearly six thousand recently built SHHA plots are attached to the sewer. The intention is to install flush toilets in old SHHA housing as well: "... it has been decreed that all areas must receive the same services and where these are lacking ... these services will be brought in" (Mosha, 1992, p. 11).

31 In Gaborone, people generally use sullage (i.e. washing water) for irrigation or to keep down yard dust or they pour it into storm drains, and they may continue these habits even after they are connected to the sewer.

32 I.e. 7.5 m^3/mo = 50 l/c/d * 5 people/family * 30 days/month * .001 m^3/liter.

33 Calculations of this sort were first made by GOB, 1991c, pp. 2-5f.

34 Notice that the cost of treating the sewage is not mentioned here. When pit latrines are emptied, the sludge is put into the sewer, so roughly the same treatment costs are incurred no matter which system the household adopts. Notice also that differential environmental costs are not mentioned here. There is no evidence of pit latrines leaching and polluting groundwater in Gaborone, but should this begin to occur, then the net social cost of the switch to waterborne sewage is further reduced.

35 We were unable to include depreciation and interest costs on the sewerage capital in the expenditure estimates of Figure 4.6.

36 The principal private hauler is Skip Hire, which also services customer-oriented businesses that want more frequent, regular, or dependable pickup than the City can provide.

37 Almost twice as much solid waste is picked up per plot per day from medium-income and high-income housing as from SHHA housing (Dohrmann et al., 1991, p. 3-8). This same study also noted that 10-20% of the solid waste received "on-site disposal" — a euphemism for small-scale burning — mostly in the low-income areas (ibid., p. 3-2).

38 Again typically, the density is two to three times higher in the low-income areas than in the high-income areas (Dohrmann et al., 1991, p. 3-6).

39　The compactors are used in the high-income areas because the trash there is less dense and hence the benefits of compaction are greater. Also, in the low-income areas, where wood fires are much used, burning ashes get into the trash barrels and can do serious damage to the hydraulics of the compactors. We had hoped to do a comparative-cost analysis on the choice between compactor and traditional trucks but were unable to uncover sufficient data. One recent study, commissioned by USAID (which, remember, wants to sell big U.S. compactor trucks) found the compactor to be "the most economical" and "the most efficient" vehicle, although it beat traditional trucks by only $51.61 to $52.38 per ton of waste collected (Dohrmann et al., 1991, pp. 3-7ff). And even that close call required a number of assumptions that are generous to compactors: 1) compactors do some 75% more plots per day because they make fewer trips to and from the landfill (though the landfill is barely out of the City); 2) compactors cost only 80% more than traditional trucks ($123 thousand versus $67 thousand); 3) both kinds of trucks have the same life — five years — though the average age of the current truck fleet is much higher than that; 4) both kinds of trucks have the same maintenance costs — some $3-4 thousand per year; and 5) those data that come from observing existing routes were not affected by the fact that compactors now do one kind of route (high-income areas)　and traditional trucks another (low-income areas).

40　The name of the company may seem curious for a solid-waste operation. The company began with the renting and servicing of portable toilets, largely at construction sites, but increasingly to varied consumers, like the army and party-givers. From this, it added septic tank cleaning and, when the GCC decided on the privatization pilot program, solid waste collection. Incidentally, the GCC retained the task of street-cleaning and litter-picking in the Daisy Loo collection areas.

41　Daisy Loo's big compactors, built in the United States and USAID-financed, do not carry the skips to the landfill but have the ability to tilt the skip up into the compactor at the back of the truck. (Actually, Daisy Loo has three big compactors, but one is usually held in reserve for emergency or replacement during maintenance.)

42　The Daisy Loo compactors require a crew of eight when they do curbside pickup, but only a crew of two when they work skips. Before concluding that skip-loading is less labor-intensive, however, one must remember all the handcart operators bringing the refuse to the skips.

43 It is not quite correct to say that Daisy Loo was "forced" to accept the trucks. It had a choice of being paid more or accepting the loan of these trucks from the GCC. But Daisy Loo did not have access to the kind of capital needed to buy its own equipment.

44 It roughly doubles the pickup time of the labor and vehicles involved, although it does not much increase the time spent going to and from the landfill.

45 The GCC Health Dept currently employs 150 people in refuse collection and 214 in litter picking and street cleaning (Dohrmann et al., 1991, p. 3-4).

46 Tires are too "elastic" to landfill, and there is no current recycling procedure for them (COG, 1993, p. 4-8). The pile would be growing except that people come to take them for various personal uses. Similarly, the tree cuttings disappear for firewood.

47 This total cost involves 28% of the City transport budget, 35% of the refuse collection budget, plus an allowance for depreciation and interest on the vehicle fleet. The 28% and 35% figures are the estimates Dohrmann received from City officials. We have recalculated Dohrmann's equipment costs to take account of inflation, which he did not, but we were unable to get a time series on the vehicle fleet and hence assumed that, in each year throughout 1982-94, depreciation was 10% of the 1994 replacement cost of the 1994 fleet and interest was 8% of half of that replacement cost (half because the average vehicle in the fleet is half worn out).

48 The only other estimate of the City's solid waste costs we know of is $1.32 per capita per year, but neither source nor method is given (Tevera, 1994, p. 24).

49 See Tevera, 1994, for a fascinating and detailed study of the socioeconomic characteristics of the scavengers and their families (at the now closed dump).

50 There are also scavengers outside the landfills, chiefly seeking cans, but they collect a much smaller percentage of the total than those at the landfill.

51 It also distributes burlap bags for free to potential collectors. The difference between the delivered-price and the collected-price has encouraged middlemen with trucks to buy from scavengers, in Gaborone and elsewhere (*Midweek Sun*, 1995, p. 10).

52 Tevera, 1994, offers another reason for the paucity of recycling in Botswana: the "abundance" of foreign exchange (p. 29).

53   They maintain that, with a proper sorting operation, the City could recover three tons of glass, four tons of tin, six tons of paper, and two tons of plastic per day, and that three to five tons per day could be separated and composted (Dohrmann et al., 1991, p. 4-9). Some glass is already being recaptured by Skip Hire at shopping centers (Gould et al., 1995, p. 10).

54   Their exact figures: Pula 32.28 per ton for "alternative 1" *versus* Pula 35.03 for "alternative 3". In 1994 US dollars, that is $19.69 *versus* $21.37.

55   The details of the annualized cost calculation are given below (converted into thousands of 1994 US dollars):

Cost Category	Alt. 1	Alt. 3
Annual Net Recurrent Cost	280.8	280.5
Landfill Construction Cost	1489.6	1489.6
All Other Capital Cost	577.1	2457.2
Life of the Landfill (years)	27	44

In short, recycling revenue barely covers the additional gross recurrent cost of the MRF, and the capital cost of the MRF is nearly two million dollars (i.e. $2457.2 - 577.1 thousands).

56   In 1994 US dollars, the annualized cost of alternative (1) is $428,000 versus $576,000 for alternative (3).

57   The high cost of water, especially during droughts, may have hastened the switch (Gould et al., 1995, p. 10).

58   See Porter, 1978b, and Porter, 1983, for discussions of the cost of mandatory deposits in the United States. While retailers dislike the returnable system everywhere, Gaborone retailers seem to have gone a step further than most in pursuing this preference, often refusing to accept empties, requiring whole cases, demanding proof that the bottles were bought in that store, giving chits rather than cash to insure that the deposits are spent there, etc. (Sandy, 1994, p. 16). One survey showed that the average retailer paid out in fact only about 80% of the bottle's official deposit value (ibid., p. 2; Somarelang, 1995, p. 7). Games such as this would quickly lose customers for the retailer in most places, but again, sparse population and great distances, coupled with the relative immobility of poor people, may give retailers local monopoly power in Botswana.

5 Comparisons and lessons

Introduction

These three cities were not randomly selected from the capitals of the Sub-Saharan African (SSA) nations. The three all are, for example, Anglophone by colonial heritage. But the cities were also not chosen specifically because their income per capita has grown particularly rapidly in recent decades, at least by the usual indicators of growth. Two of the three (Ghana and Zimbabwe) have experienced *negative* growth in GDP per capita over the last few decades, and the third (Botswana) has grown the *most rapidly* of all SSA countries.

What these three countries do share, importantly for this study, are policymakers with earnest concern for the provision of an adequate urban environment to the citizens of these cities. By almost any statistical measure of urban services, all three of these cities will appear near the top of SSA cities. Contrast the urban environments of these cities with the following assessment, that typically in SSA the level of urban services

> has deteriorated .... With rapid urbanization, many ... areas have developed with no water supply .... [Services cover] only the colonial core of the cities. (Wekwete, 1992, pp. 129f)

Since Parts 2 through 4 of this monograph have often been critical of various aspects of the provisions and the policies of the cities and their governing bodies, it is very important for readers to remember that these cities have basically provided shining examples of effort — and in many ways shining examples of achievement as well. They have all succeeded in providing an amazingly high standard of urban amenities, compared to the average of developing country cities, and not just for the well-off households.[1]

105

With that as background, we launch into a comparison of the cities, emphasizing the differences between them, in achievements and in policies, in order to draw lessons from them for other cities of the developing world that are concerned to improve their urban environments.[2]

Drinking water

There are big differences in the volume as well as the variety of urban services provided in these three cities. There is perhaps no better way to see these differences than in the water consumption data. For these three cities, the overall water consumption — for all commercial, industrial, and residential purposes in 1994 — was:[3]

Accra	86	l/c/d,
Harare	257	l/c/d,
Gaborone	174	l/c/d.

These numbers are surprising, if one thinks only of the supply side — the country where water is not considered extremely scarce supplies its capital-city citizens with far less water than the other two cities.

From the demand side, however, the water consumption figures make much more sense. Income per capita is higher in Zimbabwe, and much higher in Botswana, than in Ghana, and the demand for water of course has positive income elasticity.[4] Prices also help to explain the differences in water consumption. In Table 5.1 are shown the average monthly water bills in each of the cities. Water is sold at a lower average price in Harare than in Accra, and it is sold at a three to six times higher average price in Gaborone than in Accra. Price elasticity of demand also contributes to the consumption outcomes.

In fact, a price elasticity of demand of around (minus) one and an income elasticity of demand of around (plus) one explains almost all the observed differences in average consumption of water per capita among the three cities.[5]

Also seen in Table 5.1 is the fact that all three cities use stepped-up, "lifeline" pricing for water — i.e. prices that rise with the amount consumed. But the rates at which the prices are stepped are very different. In Accra and Harare, families that consumed 50 m^3/month in December 1994 paid about ten times as large a water bill as families that consumed only 10 m^3/month, and families that consumed 100 m^3/month paid nearly 30 times as large a

Table 5.1
Monthly household water bills

US$ December 1994

Water use	July 1984			December 1994		
	Accra	Harare	Gaborone	Accra	Harare	Gaborone
10 m³/mo	0.62	1.18	3.54	0.87	0.80	3.16
50 m³/mo	14.43	8.38	45.11	9.31	7.14	60.64
100 m³/mo	40.37	18.63	124.18	24.39	20.24	142.48

Sources: For Accra, Figure 2.2; for Harare, Figure 3.2; and for Gaborone, Figure 4.3.

water bill as families that consumed only 10 m^3/month. In Gaborone, on the other hand, the stepping-up was much more steep — families that consumed 50 m^3/month paid nearly 20 times as much as families that consumed only 10 m^3/month, and families that consumed 100 m^3/month paid nearly 50 times as much as families that consumed only 10 m^3/month. It should not surprise visitors to discover that in Gaborone only the very, very rich have lawns around their houses, that in Harare even middle-income families have lawns, and that the paucity of lawns in Accra is more due to the relative paucity of rich families than to a high water price.

Despite the stepped-up "lifeline" pricing structure in each of the cities, there is great inequality in each in the amounts of water consumed by people of different income levels. In each of the cities, the low-income families consume less than 100 liters per capita per day (l/c/d), and many in Accra and Harare less than 50 l/c/d. And many high-income families consume over 300 l/c/d, especially in Harare. This is partly due to income elasticity of demand and partly due to the way water is delivered.

Drinking water is delivered in each of these cities primarily either through in-house (and in-yard) taps or through public standpipes — see Table 5.2. Typically, in developing country cities, the rich are willing to pay the higher costs of in-house delivery, and the poor prefer the cheaper — and often subsidized or even free — standpipe water. It is therefore surprising to find that only in the wealthiest of the three cities (Gaborone) are public standpipes heavily relied on as a means of water delivery. In Accra, only about 10% of the population is served by communal standpipe, and in Harare, almost no public standpipes are provided. In Gaborone, on the other hand, roughly half the houses consume water from standpipes.

Gaborone's standpipe users are presumably happy with their inferior water accommodation since almost all of them have foregone the (usually available) opportunity to upgrade (at a price) to in-house taps. The difference between Gaborone and Accra is that Gaborone's standpipes are many and operational, whereas Accra's standpipes are few and badly maintained.[6] The difference with Harare is that very poor Harare families do not receive the option of standpipes and thus may be being forced into a consumption pattern beyond their incomes.

The three cities add little evidence to the debate whether public standpipes should be individually monitored and the water priced. Accra is beginning to staff some of its standpipes, but it is still too early to know if their operation will improve. Gaborone, with standpipes unstaffed and water unpriced, has kept up an excellent record of maintenance and efficiency, proving that unmonitored standpipes need not leak, break down, or waste water.

Table 5.2
Water source by income group

	Accra	Harare	Gaborone
In-house or in-yard tap	about two thirds	almost all	about half
Public standpipe	about one tenth	almost no one	about half
Other	about one fourth	about one tenth	almost no one

Note: "Other" means wells, vendors, neighbors, etc.

The water agencies of the three cities also display very different revenues and expenditures. Table 5.3 shows water income and expenditure per capita, and Table 5.4 shows water income and expenditure per cubic meter. Consider first the costs, and focus on the 1994 average cost per cubic meter. Two things stand out:

1   The average cost of water is much higher in Gaborone, and it has nearly doubled (in real terms) in the past decade. Gaborone is wise to discourage frivolous water consumption by means of its high charges to heavy users.

2   The average cost of water is comparable in Accra and Harare, and in both cities the average cost is much lower than in Gaborone. But the provision of water is not at all comparable in Accra and Harare, either in the quantity provided (per capita) or in the quality provided.[7] This tells us something about economies of scale in water production and distribution (Harare produces some four times as much as Accra) or about relative inefficiency in the Accra water operation (or both).

As one might expect, the total revenue of the water utility, either per capita or per cubic meter, is much higher in Gaborone than in Accra or Harare. More interesting — and more surprising in SSA where parastatals have become notorious for failure to cover even operating costs — all three of these water agencies cover their operating costs. Depending on how credible are our additions for interest and depreciation costs, perhaps only Accra covers the *full* costs of its water production and distribution, but even Accra has only been achieving this relatively recently. In an era when central governments are less prone to subsidize urban drinking water investments, it is increasingly essential to generate incomes that cover the full costs, including the capital costs, of water production.

In the treatment and distribution of water, some is always lost. But these losses — called unaccounted-for water (UFW) — are very low in Harare and Gaborone, perhaps less than 15%, with a collection rate of nearly 100% of billings. In Accra, at the other extreme, UFW may run as high as 50%; the collection rate has only recently been raised to around 90%, having been only 50% in the mid-1980s. UFW represents wasted resources if the water is truly lost, but even if it is just diverted surreptitiously or goes unpaid-for, this represents unintended income redistribution and adds to the budgetary problems of the water agency.

Table 5.3
Water utility income and expenditure per capita

US$ 1994 per capita

	1983			*1994*		
	Accra	Harare	Gaborone	Accra[2]	Harare	Gaborone
Income	0.42	13.98	58.45	10.08	13.87	79.04
Expenditure						
exc i+d[1]	0.89	11.73	30.71	4.24	9.78	29.73
inc i+d	0.95	19.84	78.51	4.81	18.47	95.29

Notes

1   exc i+d means excluding a charge for interest and depreciation; inc i+d means including imputed interest and depreciation charges.
2   The Accra figures are for 1993, not 1994.

Sources: Water utility accounts (for Accra, Figure 2.1; for Harare, Figure 3.4; and for Gaborone, Figure 4.2).

Table 5.4
Water utility income and expenditure per cubic meter[1]

US$ 1994 per m^3

	1983			1994		
	Accra[2]	Harare[2]	Gaborone	Accra[2]	Harare	Gaborone
Income	0.22	0.14	0.55	0.27	0.15	1.24
Expenditure						
exc i+d[3]	0.10	0.16	0.29	0.12	0.10	0.47
inc i+d	0.11	0.28	0.74	0.14	0.20	1.50

Notes

1   The cubic meters refer to sales in Accra and Gaborone, but to production in Harare.
2   The Accra figures are for 1986 and 1993, not 1983 and 1994; the Harare figure is for 1985, not 1983.
3   exc i+d means excluding a charge for interest and depreciation; inc i+d means including these charges.

Sources: Water utility accounts.

There are many ways in which families in these cities can get "free" water, not just through illegal connections or unpaid bills. Wells, boreholes, unbilled or unpaid government agency use, unmetered taps, broken meters, and housing where the water bill is included in the rent are all examples of water consumption where the consumers pays no, or a low, price on the margin for water. All of these encourage frivolous water consumption, and all generate horizontal inequities among water consumers. Consistent pricing of water should always be of high priority.

Human waste

The processing of human waste is done very differently in the three cities. A rough description is offered in Table 5.5. The most dramatic differences concern:

1   *Sewers.* Accra essentially has no sewer system, while both Harare and Gaborone are fully sewered in the sense that almost everyone could be connected up to the existing system. In fact, however, most of the high-income Harare households (on large, low-density plots) prefer (and are permitted) to rely on septic tanks, while most of the low-income Gaborone households (on small, high-density plots) prefer (and are permitted) to rely on in-yard pit latrines.

2   *Septic tanks.* Not the system of choice in crowded urban settings, since proper drainage of a septic tank requires a significant amount of land, septic tanks are nevertheless used extensively in the high-income, low-density areas of both Accra and Harare — but for different reasons. In Accra, the alternative of a sewer connection rarely exists, while in Harare, the high connection and sewerage costs induce those with large plots to prefer septic tanks.[8]

3   *Pan or pit latrines.* Except in a few unplanned squatter settlements, pan or pit latrines are not utilized in Harare. But they are extensively used by the low-income, high-density households of Accra and Gaborone. In Accra, the pan latrine dominates, with the nightsoil being regularly collected by "conservancy laborers" at a price and dumped into the ocean; in Gaborone, the pit latrine dominates, with the necessary periodic emptying being done by City vehicles at no price.

113

Table 5.5
Types of human waste disposal systems used

	Accra	Harare	Gaborone
Water-flushed toilets and sewer system	a few rich	most rich,	most rich, a few poor[1]
Water-flushed toilets and septic tanks	most rich	most rich, a few poor	a few rich
In-yard pit or pan latrines	about half the poor	almost no one	most poor
Public toilets	about half the poor	almost no one	almost no one
Length of sewer system[2]	0.02	4.10	1.85
Public toilet frequency[3]	1.50	0.50	0.85

Notes

1 "Rich" means low-density, high- and middle-income residents; "poor" means high-density, low-income residents.
2 Length of sewer system counts all trunk sewers and collector mains and is measured in kilometers per thousand population (1989).
3 Public toilet frequency counts all public toilet seats and is measured in seats per thousand population (1989).

Sources: For Harare and Gaborone data, Colquhoun, 1990a; and for Accra data, Engmann, 1990.

4   *Public toilets.* Only in Accra do the poor rely heavily on public toilets, and there are many more of them in Accra (per thousand people) than in Harare or Gaborone. But there are still too few for the numbers of users, and they are often unclean and overfilled.

Sewers are, of course, the ultimate answer to all sewage problems in large, densely populated cities. Harare and Gaborone are lucky to have already made the large investments required by sewer systems, so that they now need to undertake only the expense of maintenance and expansion. For Accra, given Ghana's economic straits, a full sewer system may not become feasible for decades.

But even the connection to an existing sewer system is expensive. Witness Harare, where even the well-off regularly opt for septic tanks instead. And for the poor, the costs of in-house, water-flushed, sewer-attached toilets may be a very high-cost luxury of very low priority to the family. Witness Gaborone, where the poor rarely opt for attachment to the sewer even where it is feasible.

Septic tanks, of course, cannot be even a satisfactory interim measure in densely-populated areas, but regularly emptied in-yard pit latrines or well-maintained public toilets can be. Which should be chosen depends greatly on public attitudes as well as on income levels. Gaborone, with its much higher average incomes, has been able to provide in-yard pit latrines to the households of its low-income, high-density residential areas; Accra, with its much lower average incomes, probably should attempt to improve and expand its system of public toilets.

There are externalities involved with both private pit latrines and public toilets (of whatever kind). The government must insure that adequate emptying services are available for pit latrines, and perhaps some subsidy is called for, at least in low-income areas. Subsidies are almost certainly appropriate for public toilets, to insure that they are properly cleaned and widely used.

The interesting questions involve "when" — when to switch from public toilets to in-yard pit latrines, and when to switch from in-yard pit latrines to in-house, water-flushed, sewer-attached toilets. The answer, in principle, is a simple one: when the household is willing to pay the social marginal cost of the switch. A lesson from Harare, and perhaps one soon to be learned also in Gaborone, is that forcing the low-income households onto sewer systems either requires heavy subsidy or reduces the welfare level of the low-income household.

Finally, there is the question of sewage treatment. Simply dumping sewage, on land or into an ocean, is clearly inappropriate in a crowded city environment. But advanced sewage treatment is usually not appropriate, either. Where land is plentiful and temperatures high, stabilization ponds are cheap and effective. In countries where water is scarce, advanced sewage treatment may be efficient as a means of reusing water, but one should always check first to be sure that it is cost-effective.[9]

Solid waste

The three cities generate comparable amounts of solid waste — a total of something like two thirds of a kilogram per capita per day, about two thirds of it residential; the three cities collect this waste in comparable ways — some curbside pickup and some container (or skip) emptying; and the three dispose of the waste similarly — taking it to a landfill where some scavenger recycling takes place.

While the differences between the cities are not great, they do suggest a number of solid waste policy issues and point toward policy solutions.

In each of the cities, two basic kinds of refuse collection are practised: house-to-house curbside pickup and container emptying. The former is the ultimate practice, by analogy to the current practices of cities in industrialized countries; but there are a number of reasons for supplying regularly spaced and regularly emptied containers (i.e. "skips") in the low-income, high-density parts of the cities of developing countries. First, technically, these areas generally have no, or few, or narrow streets so that large refuse-collection trucks cannot maneuver through them; curbside pickup may not become feasible until the areas are fully redeveloped. Second, if it is the city's intention to require user fees that cover costs, the willingness of poor families to pay for curbside service is too low to cover costs; any effort to provide curbside pickup, and to charge its full cost, will just lead to significant illegal dumping, defeating the whole purpose of the service. Third, the poor have a lower opportunity cost of time, and carrying waste a few meters to large containers is not a great burden.[10] And finally, by providing two kinds of refuse collection, the city can charge the well-off a price above cost and use the surplus to cross-subsidize the container service in the poorer areas. The evidence from Accra and Harare strongly urges against trying to charge the poor for solid waste collection.

In both Harare and Gaborone, twice-weekly collection is undertaken. The evidence from cities in much richer North America is that voters/residents are rarely willing to pay the extra cost of such frequent service. And health considerations do not require it, even in such hot climates. Budget reductions

116

here — and in the extent of the street-manicuring in Harare — would permit the cities to reduce the "service levies" in poor areas or to provide more highly demanded services instead.

In the contemporary policy environment, there is an excessive concern that everything should cover its own costs through user fees. This poses many problems with respect to solid waste collection. First, fees here are rarely true prices — i.e. so much money per kilogram of waste collected — but rather are thinly disguised, regressive additions to property taxes. This in itself suggests getting rid of the disguise. There is little virtue in breaking the property tax up into multifarious earmarked segments.

And second, refuse collection has many attributes of a public good — when a family's trash is properly collected and removed, it benefits the family's neighbors as well as the family itself, and it would burden the family's neighbors as well as the family if a city were to stop collecting from those who would not pay. We must remember that the optimal social price of a pure public good is zero.

Finally, solid waste collection is not equally a public good in high-density and low-density residential areas. For the poor, their high-density residential areas are essentially urban "commons" areas, overlittered in the same way and for the same reasons as rural common fields are overgrazed; here, if solid waste collection is priced, it will not be purchased since one family's purchase of proper disposal barely dents the overall litter of the surrounding common.[11] In low-density areas, on the other hand, neighbors live further away, and trash is much more a "private bad" that people will be very willing to pay to have removed.

In Accra and Gaborone, parts of the solid waste collection system are operated in the private sector. Broadly, the evidence is beginning to suggest that the private sector is able to lower costs, lower prices, and produce a service that is more valued by households. The lower costs seem to be achieved principally by the application of simple, labor-intensive technology, emphasizing handcarts and open trucks rather than complex, capital-intensive multilift and compactor vehicles.

Finally, all three of the cities have scavenger activity and unsubsidized, market-driven recycling. This should not surprise anyone since all three countries are still at the stage of development where extensive private, profit-driven recycling still occurs. In fact, such activities will probably expand in the next few decades of development; as urban population densities grow, transport costs decline, and industrialization proceeds, such private-sector recycling should increase, in both extent and intensity. The major lesson here is that the government should simply keep its hands off: public-sector recycling, with automated materials recovery facilities (MRFs) — such as those now becoming socially desirable and even privately profitable in the cities of West Europe and North America — are surely premature in Accra, Harare, and Gaborone.

Final thoughts

There is a tendency to think of the urban environment as moving linearly in the process of development from that of a small village to that of today's big cities in the industrialized world. Initially, in the small village, few services are formally provided; citizens find their water in nearby natural sources — such as shallow wells, ponds, or rivers — and they dispose haphazardly of their limited, dispersed quantities of waste. At the other extreme, publicly regulated or publicly owned giant monopolies pipe in treated water to every resident's house and pipe out all the septage and sullage for treatment. Each household's solid waste is also collected, regularly and at curbside, extensively recycled, and disposed of in an environmentally safeguarded landfill.

The end points of this transition are roughly accurate; the error is in thinking that the transition consists simply of providing state-of-the-art amenities to an ever larger portion of the citizens. Doing so means trying to provide to all citizens the advanced level of services that only the well-off are currently willing to pay for. If state-of-the-art amenities are provided to the poor, either they must be forced to accept them prematurely or the services must be subsidized. Where the poor resist, or budgets are limited, or both, the result is that an advanced level of services is provided to the rich and that little, or nothing, is provided to the poor.

This process and this outcome are observed in cities throughout the developing world. In some cases, the capital costs of state-of-the-art services are so high that politically feasible prices fail to provide sufficient revenue to cover even maintenance costs — then, not only do the poor get little or no service but the rich may get subsidized service:

> One common objective found among those in charge of designing urban policies in developing countries is the goal to make cities serve more effectively the preferences of the better-off sections of the urban population .... The failure to recoup urban services costs from beneficiaries ... has frequently led to extensive subsidies, particularly to higher-income groups. (Linn, 1983, pp. 28, 80)

The failure to provide an appropriate — i.e. adequate *and* affordable — kind of service to the urban poor often dooms them to an intolerably unhealthy urban environment. The services we have been discussing are not like food and clothing; water provision and waste disposal are not well handled by private markets. Handcart or tanker-truck vendors, for example, are generally the most costly way to acquire water; and unsubsidized private operators are rarely able to profit from waste disposal. The government's failure to provide "second-best" service to the poor usually means that the poor will get expensive "third-best" service, or none at all.

For almost every urban service, there is a "second-best" provision that is either affordable by the poor themselves if the service is a private good, or affordable by the municipal budget if there are strong elements of public good or externality involved. For drinking water, public standpipes; for human waste, pit-latrine construction or cleaning services and well-maintained public toilets; for solid waste, regularly spaced "skips" to which refuse can be delivered.[12]

The question then arises, how does the city decide who should get which service? Often, it is better if the city does not decide this. The efficient provision (by a budget-constrained city) may require offering residents a choice between two kinds of services and letting them self-select the one they are willing to pay for (Fischer and Porter, 1993). Forcing poor residents to accept and pay for "first-best" service may be just as inefficient as forcing "third-best" service on them owing to the city's failure to provide a "second-best" alternative.

Not only must "appropriate" services be offered, but they must be produced by "appropriate" technologies. One of the few advantages of being a city of largely poor people is that wages are low and hence that urban services can be cheaply provided if they are labor-intensive. For each of the services we have been discussing, the "second-best" delivery system is more labor-intensive: the deployment of operators and maintenance workers at public standpipes and public toilets; the use of handcarts and open trucks for solid waste collection.[13]

In short, the provision of an appropriate urban environment — adequate and healthful for all residents — is not beyond the budgets of developing-country cities, even when one recognizes that it is neither possible nor desirable to charge cost-covering fees for many of the services provided. It requires greater reliance on simple services, greater attention to low capital cost, and greater concern for proper maintenance.

In the United States about 12 per cent of gross national product, or something like US$2500 per person, is spent annually on health. This is in excess of the per capita national income of several countries labeled "upper middle income" [by the World Bank] .... And yet, in the late 1970s citizens in China and Sri Lanka, two countries that were (and are) very clearly "poor," enjoyed a life expectancy at birth only about 6 or 7 years less than the United States .... A key reason behind this achievement has been the availability of sanitation facilities and public health care for most members of society. In addition to funds being made available, their delivery systems were reliable .... Relative to what it can achieve, it does not require enormous capital expenditure. (Partha Dasgupta, 1993, pp. 92f)

Notes

1 These three cities compare favorably in this respect not only with other SSA cities but also with many Asian cities, even in those countries that have experienced rapid growth (Coolidge et al., 1993). In Jakarta (Indonesia), for example, GDP per capita has grown dramatically over the past 30 years, but the poor there suffer a far worse urban environment than in any of these three cities (Porter, 1996).

2 The comparisons of the three cities are not as quantitative as we had hoped. Although a great deal of comparative data, both physical and financial, was collected in 1989 for Harare and Gaborone (Colquhoun, 1990a), and the very same data were collected for the same year, presumably comparably, for Accra (Engmann, 1990), we have made little use of them here, partly because much of the Gaborone data were not ascertained and partly because we were unable to reconcile many discrepancies with the relevant agency accounts.

3 l/c/d means liters per capita per day. The water consumption data are from the relevant agency accounts; the population figures are my extrapolations from the most recent censuses. Throughout this Part 5, whenever the source of information is not given, the relevant data can be found in Parts 2-4 of this monograph.

4 Income per capita is roughly twice as high in Zimbabwe as in Ghana and five times as high in Botswana as in Ghana. See Table 1.1 in Part 1.

5 Consider the formula that compares any pair of the three cities:

$$D_{c/n} = E_y D_{y/n} + E_p D_p ,$$

where $D_{c/n}$ is the percentage difference between the cities in water consumption (c) per capita (n), $D_{y/n}$ is the percentage difference between the cities in income (y) per capita (n), D_p is the percentage difference between the cities in average water price (p), E_y is the income-elasticity of demand (assumed the same in both cities), and E_p is the price-elasticity of demand (also assumed the same in both cities). The numbers given in the text are roughly consistent with this formula with $E_y = 1$ and $E_p = -1$.

6 There have never been more than 350 public standpipes in all of Accra, and there were as few as 100 a decade ago. Thus, something like one fourth of Accra's population relies on shallow wells, vendors, neighbors, or polluted natural water sources — the absence of the "second-best" delivery system (public standpipes) forces the poor there to adopt even lower-quality or higher-priced "third-best" delivery systems.

7 The quality difference resides in the frequent supply interruptions and pressure variations in Accra, both of which are rare in Harare. Tap water in both cities is safe to drink.

8 Those with small plots might also so prefer, but the City does not permit the exercise of this preference unless the plot is large.

9 Direct reuse of wastewater is limited to industrial applications even in highly developed countries, although indirect reuse is practiced — by the time the Mississippi River (USA) has run its course, every drop has been drunk many times (Ray, 1995, p. 307).

10 Recall that waste generation has positive income-elasticity, so that poor people generate less waste that needs carrying.

11 Litter is essentially a public "bad" rather than a public good. Just as the "free market" underproduces public goods, it overproduces public bads.

12 For each of these, there is a distinctly inferior "third-best" service. For drinking water, private tanker or handcart vendors or polluted ponds or streams; for human waste and solid waste, the "commons" areas — that is, streets, fields, ditches, and storm drains.

13 This is not to say that all labor-intensive techniques are good and all capital-intensive techniques are bad. Handcart vending of drinking water, for example, is very labor-intensive *and* very inefficient. The argument here is rather that state-of-the-art techniques are — by definition, since they come from industrialized, high-wage countries — highly capital-intensive, and that careful benefit-cost analysis should be applied before they are adopted.

6 Appendices

Appendix A: People who helped

Accra

Abakah, Moses. Accountant. Waste Management Division. Accra Metropolitan Authority.

Amankwah, E. O. Department of Town and Country Planning. Ministry of Environment.

Awaitey, James F. T. Accra Metropolitan Budget Officer. Ministry of Finance and Economic Planning.

Awoonor-Williams, M. K. Senior Statistician. Ghana Statistical Service.

Awuku, Martin K. Head Commercial Officer. Ghana Water and Sewerage Corporation.

Botchie, George. Senior Research Fellow. Institute of Social, Statistical and Economic Research. University of Ghana.

Dennis, K. E. Assistant Chief Executive Officer. Accra Metropolitan Authority.

Digby, Peter K. W. Statistical Advisor. Ghana Living Standards Survey. Ghana Statistical Service.

Ewool, Godfrey. Project Officer (Infrastructure). Resident Mission in Ghana. World Bank.

Gough, Kate. Department of Geography. University of Copenhagen.

Koch, Helmut R. Project Manager/Financial Advisor. Waste Management Division. Accra Metropolitan Authority.

Kwa-Eshun, Anthony. Acting Chief Accountant. Ghana Water and Sewerage Corporation.

Nai, G. Director for Water. Ministry of Works and Housing.

Nartey-Tokoli, I. B. Planning Officer. Waste Management Department. Accra Metropolitan Authority.

Quarcoo, E. C. Private Secretary. Ghana Water and Sewerage Corporation.

Tutu, K. A. Lecturer. Department of Economics. University of Ghana.

Nairobi

Adrian, Jean-Christophe. Advisor. Sustainable Cities Program. United Nations Center for Human Settlements (Habitat).

Alabaster, Graham P. Human Settlements Officer. Research and Development Division. United Nations Center for Human Settlements (Habitat).

Dzikus, Andre. Human Settlements Officer. Research and Development Division. United Nations Center for Human Settlements (Habitat).

Lorentzen, Jens. Human Settlements Advisor. Urban Management. Technical Co-Operation Division. United Nations Center for Human Settlements (Habitat).

Mutizwa-Mangiza, Naison D. Human Settlements Officer. Research and Development Division. United Nations Center for Human Settlements (Habitat).

Wekwete, Kadmiel. Urban Management Advisor. Urban Poverty Research. United Nations Center for Human Settlements (Habitat).

Harare

Bowling-Scott, D. Chair. Environment 2000. Water Conservation Committee. City of Harare.

Brushett, Stephen J. Deputy Resident Representative. Resident Mission in Zimbabwe. World Bank.

Chenga, Matthias B. Associate. Brian Colquhoun, Hugh O'Donnell & Partners. Harare.

Chikungwa, Miriam. Accountant. Central Accounts Division. City Treasurer's Department. City of Harare.

Davies, Robert. Lecturer. Department of Economics. University of Zimbabwe.

Davison, Celia. Lecturer. Department of Rural and Urban Planning. University of Zimbabwe.

Dickens, Graham. Manager. The Bronte Hotel. Harare.

Gwasira, Brian. Cleansing Inspector. Department of Works. City of Harare.

Hewat, Charlene. Secretary General. Environment 2000.

Hicks, R. W. Associate. Brian Colquhoun, Hugh O'Donnell & Partners. Harare.

Jenner, Ronald. General Manager. Art Reclamation.

Mahachi, Tongai S. Director of Works. City of Harare.

Makwembere, R. Accounting Officer. Revenue Division. City Treasurer's Department. City of Harare.

Maya, R. The Southern Centre for Energy and Development. Harare.

Meyer, D. Former Chair. Water Conservation Committee. MD Surgimed.

Mkudu, George. Deputy Engineer. Water and Sewerage Division. Department of Works. City of Harare.

Mubvami, Takawira. Lecturer. Environmental Surveys and Housing. Department of Rural and Urban Planning. University of Zimbabwe.

Mumbengegwi, Clever. Chair. Department of Economics. University of Zimbabwe.

Mutuwa, Stanley. Principal Engineer. Water and Sewerage Division. Department of Works. City of Harare.

Ndlovu, L. B. Chair and Lecturer. Department of Rural and Urban Planning. University of Zimbabwe.

Paruwani, J. Senior Clerical Officer. Revenue Division. City Treasurer's Department. City of Harare.

Rowett, Felicity J. General Manager. National Waste Collections.

Sithole, V. Chief Engineer for Water and Sewerage. Department of Works. City of Harare.

Tevera, D. S. Chair and Lecturer. Department of Geography. University of Zimbabwe.

Zata, Myles C. T. Amenities Manager. Department of Works. City of Harare.

Zhungu, D. T. Accountant. Central Accounts Division. City Treasurer's Department. City of Harare.

Gaborone

Brahmbhatt, Pushkar A. Project Development Officer. U.S. Agency for International Development (AID).

Davies, Gerald. Owner/Manager. Skip Hire (Pty) Ltd.

Dijeng, Gerald M. Managing Director. DaisyLoo (Botswana).

Gould, John E. Senior Lecturer. Department of Environmental Science. University of Botswana.

Jeter, Howard. U.S. Ambassador to Botswana.

Khupe, John S. N. Senior Environmental Engineer. Gaborone City Council.

McDonald, Ian. Director. Waste Paper Recovery Botswana. Pyramid Holdings (Pty) Ltd.

Masenya, Francis. Chief Environmental Health Officer. Gaborone City Council.

Matsheka, Casper. Assistant Manager. Metal Box Botswana.

Moroka, L. K. Planning Engineer. Water Utilities Corporation (WUC).

Mosha, Aloysius C. Senior Lecturer. Department of Environmental Science. University of Botswana.

Nels, Christian. Project Coordinator. Waste Management Project. German Technical Cooperation.

Ookeditse, S. Manager of Operations. Water Utilities Corporation. Gaborone.

Rae, John. Manager. Cresta Lodge. Gaborone.

Rao, K. N. Senior Assistant Librarian. Special Collections. University of Botswana Library.

Rathedi, M. Lecturer. Department of Economics. University of Botswana.

Sahai, R. Chief Technical Officer. Sanitation. Gaborone City Council.

Seabelo, Solomon. Ward Officer. Naledi. Self Help Housing Authority. Gaborone.

Sefe, F. T. K. Senior Lecturer. Department of Environmental Science. University of Botswana.

Segosebe, E. M. Lecturer. Department of Environmental Science. University of Botswana.

Selotlegeng, Kodise. Senior Public Health Engineer. Project Waste Management/Protection of Water Resources.

Silitshena, R. M. K. Chair. Department of Environmental Science. University of Botswana.

Washington

Bartone, Carl. Principal Environmental Specialist. World Bank.

Bernstein, Janice. World Bank.

Blaxall, John. Manager. Transportation, Water & Urban Development Department. World Bank.

Carroll, Alan. Senior Urban Specialist. Infrastructure Operations Division. Western Africa Department. World Bank.

Coolidge, Jacqueline G. World Bank.

Crowley, Diana. Program Director. The Environmental and Natural Resources Policy and Training Project. MUCIA.

Garn, Harvey A. Water and Sanitation Division. Transportation, Water & Urban Development Department. World Bank.

Hafner, Craig R. Deputy Director. Water and Wastes Management. Environmental Health Project. WASH.

Obeng, Letitia A. Senior Water and Sanitation Specialist. AFTES Division. Africa Technical Department. World Bank.

Roome, John A. Senior Financial Analyst. Infrastructure Operations Division. Southern Africa Department. World Bank.

Shepherd, K. John. Senior Water Resources Management Specialist. Infrastructure Operations Division. Southern Africa Department. World Bank.

Wright, James. Division Chief. Infrastructure Operations Division. Western Africa Department. World Bank.

Ann Arbor

Bulkley, Jonathan W. Professor. Civil and Environmental Engineering. The University of Michigan.

Eichmann, Richard. Undergraduate. The University of Michigan.

Greimel, Timothy. Undergraduate. The University of Michigan.

Keselman, Eugene. Undergraduate. The University of Michigan.

Short, Tonia S. School of Public Policy. The University of Michigan.

Appendix B: Capital stock estimates

Estimates of the real capital stock (in US$ 1994) are used at various places in the text. These estimates should be recognized as being very rough, being calculated as follows.

In the cities under review, all that is reliably known about the capital stock involved in the services we are studying is usually the gross investment for each of several recent years. Depreciation figures are also available for recent years in some City accounts, but these are almost certainly not intended as measures of the real rate of economic deterioration.

If we assume that always, before some year, say 1983, net real investment grew at some rate, g, per year and the real capital stock depreciated at some rate, d, per year, then we can use the gross real investment figure for 1983 (I_{83}) to estimate the real capital stock in 1983 (K_{83}):

$$K_{83} = I_{83}/(g + d). \hspace{3cm} [\text{B-1}]$$

In each year subsequent to 1983, the real capital stock estimate is simply the real capital stock estimate of the previous year, minus depreciation at rate d, plus the gross real investment of that year. The values of g and d are arbitrarily set at .06 and .02, respectively.

The initial year's real capital estimate is particularly sensitive to the real gross investment figure used for the initial year. Since these investment figures fluctuate a great deal over the years for which there are data (see Table B.1), we have used the trend value for 1983 in place of the actual value.[1]

Some check on our water capital estimates can be made through examination of the implied average capital-output ratio (ACOR) and the implied incremental capital-output ratio (ICOR) since the ACOR incorporates the uncertain initial capital stock estimate (K_{83}) while the ICOR utilizes only the more confidently known investment data after 1983. Figure B.1 displays these ACOR and ICOR estimates for each of the three cities.[2] Note that the ACOR and ICOR estimates are quite close to each other for both Harare and Gaborone, but that the ICOR estimate is some four times larger than the average of the ACORs for Accra. This strongly suggests that the initial capital stock (K_{83}) was greatly underestimated — i.e. that Accra still has sizeable water capital remaining from investments made long ago. When the interest and depreciation costs of the capital are included in the expenditure calculations of the text, we have charged depreciation (at .02 per annum) and interest (at .08 per annum) to our estimate of the capital stock.[3]

127

Table B.1
Recent real gross water investment in the three cities

(in US$ 1994 millions)

Year	Accra	Harare	Gaborone
1983	0.007	5.785	0.459
1984	0.037	6.895	5.431
1985	0.079	9.387	40.869
1986	1.770	8.332	5.830
1987	0.112	1.374	3.759
1988	3.446	1.511	—
1989	0.133	2.254	0.184
1990	0.045	10.334	0.581
1991	0.450	8.578	12.549
1992	0.872	5.224	51.120
1993	1.289	6.500	6.737
1994	na	5.008	4.754
Regression Trend Estimate for 1983	0.041	4.920	2.219

1   na means not available.
2   — means zero.
3   See text for explanation of regression.

Sources: City accounts.

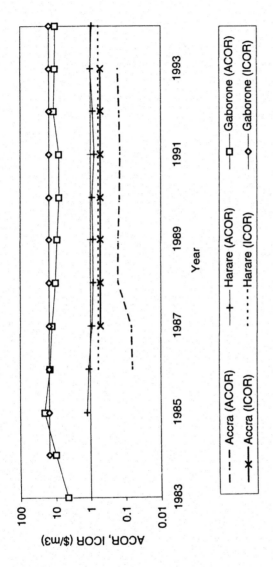

Figure B.1. Water capital-output ratios (US$ 1994 per m³, log scale)

129

Notes

1   The trend value for I_{83}, also given in Table B.1, is derived from an ordinary least-squares regression of the (natural) log of investment on time.

2   The formulas for each: $ACOR_t = K_t/Q_t$ and $ICOR = (K_{94} - K_0)/(Q_{94} - Q_0)$, where K_t is the estimated capital stock in year t, Q_t is the actual water production (or sales) in year t, and the subscript, 0, refers to the earliest year for which we could find production (or sales) data.

3   See Appendix C for an explanation of where and why interest and depreciation charges should be included.

Appendix C: Marginal-cost pricing

It has been an axiom of economists for more than a century that an efficient allocation is achieved only if the price of a good (P) equals the marginal social cost (MSC) of providing that good. This is little more than common sense. A person who is willing to pay a price of P shows that he or she is willing to give up P dollars worth of other consumption. But it requires MSC dollars worth of resources to produce one more unit of the good in question. Thus, as long as P > MSC, the consumer is willing to pay enough in foregone resources elsewhere to produce another unit. So welfare can always be increased — in the sense that an additional consumer can be made better off without making anyone else worse off — by lowering P until it just equals MSC.

The MSC in the above paragraph is of course the short-run MSC, with the overhead costs of the capital investment being previously committed, locked in place, and without opportunity elsewhere.

Nevertheless, at several places in the text, big, indivisible, durable investments are considered, and it is useful to try to attribute the costs of these investments to their users since, before the investment is made, the MSC includes these investment costs. The concept "long-run marginal cost" has come into being as a way of making this attribution. This appendix looks more closely at this concept and the sense in which the resulting attribution is useful.

Technically, an addition of one unit to output today might have a huge marginal cost — if it required a big investment that lasted a long time simply to make it feasible. The concept of "long-run marginal cost" has been developed in order to attribute to any increment in output in any particular year only the average cost of the capital investment over time when a big, indivisible, durable investment must be undertaken in order to produce additional output.

Consider, for example, an investment that is made now (t = 0), that costs C (in real terms), that wears out gradually over the ensuing T years, and that in itself (for simplicity we ignore other factors of production) produces Q units of output each year. For simplicity of the algebra, let us assume that $(1/T)^{th}$ of the initial investment wears out each ensuing year and is replaced each year, so that the initial capital (C) is maintained forever at its initial level.

The present value of the cost of such an investment (PV_1), made now (with an expenditure of C in t = 0) and maintained every year thereafter to infinity (with an expenditure of C/T in t = 1, 2, ...) is

$$PV_1 = C\left[1 + \frac{1/T}{(1+i)^1} + \frac{1/T}{(1+i)^2} + \Lambda\right], \qquad [C\text{-}1]$$

131

where i is the relevant (real) discount rate. Summing this series, we can write the PV_1 more succinctly:

$$PV_1 = C\left[1 + \frac{1}{iT}\right].$$ [C-2]

The present value of the total cost of the output that is produced by this investment (PV_2) can also be written as

$$PV_2 = xQ[(1+i)^{-1} + (1+i)^{-2} \ldots],$$ [C-3]

where x is the marginal capital cost (and also the average capital cost) of each unit of output, so that the product, xQ, is the total cost of output in each year (and the output is assumed to begin to flow in the year after the initial investment of C is first undertaken). This series [C-3] may also be summed:

$$PV_2 = \frac{xQ}{i}.$$ [C-4]

PV_1 and PV_2 are simply two ways of expressing the present value of the total cost of the investment. They are identical. Indeed, the fact that they are identical is what lets us calculate the value of x. Setting PV_1 in equation [C-2] equal to PV_2 in equation [C-4] permits us to derive the long-run marginal (and average) capital cost of each unit of output, x, as

$$x = [C/Q][i + 1/T].$$ [C-5]

Equation [C-5], the "long-run marginal cost" of the capital, x, has a nice interpretation. *It consists of two parts, the interest cost on the investment per unit of output, iC/Q, plus the depreciation of the investment per unit of output, C/(QT).* Both of these two components of cost are included in the calculations of the text whenever the total cost of an indivisible, durable investment is converted into cost per unit of output. Numerically, we have always assumed that i = 8% per annum and T = 50, so that the annual interest and depreciation cost of capital in the text is 10% of the real value of that capital.

Appendix D: About the authors

Richard C. Porter

Professor in the Department of Economics, the University of Michigan, with Ph. D. from Yale University, 1957. Undertook the fieldwork of this research on an African Regional Research Grant from the Fulbright Program. Specializes in environmental and development economics. Has lived in India, Pakistan, Colombia, Kenya, Puerto Rico, Indonesia, South Africa, Ghana, Zimbabwe, and Botswana, advising, researching, and/or teaching. Is 65, is still married, and enjoys baseball, backpacking, scuba diving, and involvement with non-Western cultures.

Louis Boakye-Yiadom Jr.

Lecturer in the Department of Economics, University of Ghana, with M. Phil (Economics) from the University of Ghana, 1993. Specializes in public finance and economic theory. Has participated in studies on financial markets in Ghana (sponsored by USAID) and on environmental-economic linkages. Is 31, is recently married, and enjoys gospel music, reading, draughts, and cards.

Albert Mafusire

Lecturer in the Department of Economics, University of Zimbabwe, with M.Sc. (Economics) from the University of Zimbabwe, 1992, and M.A. (International Economics) from the University of Sussex, 1994. Specializes in foreign trade and foreign investment. Has participated in studies on enterprise development in Zimbabwe (sponsored by the World Bank). Is 29, is recently married, plays the trombone, and enjoys gospel music, rural life, and farming.

B. Oupa Tsheko

Lecturer in the Department of Economics, University of Botswana, with M.A. (Economics) from the University of Manitoba, 1994. Specializes in international economics and macroeconomics. Is currently conducting research on water and waste in Francistown (Botswana). Is 30, is single, and enjoys participation in church activities.

Bibliography

Aboagye, I. (1995), 'Garbage Drowns Accra ... As AMA Faces Cash Problems', [Accra] *Daily Graphic*, 15 February.

Abugri, G.S. (1995), 'The Flood Disaster', [Accra] *Daily Graphic*, 7 July.

Akuffo, S.B. (1989), *Pollution Control in a Developing Economy: A Study of the Situation in Ghana*, Ghana Universities Press.

Amoah, A.E. (1995), 'GWSC to Privatise Services', [Accra] *Daily Graphic*, 7 April.

Amuzu, A.T., and J. Leitmann (1991), *Environmental Profile of Accra*, Urban Management and Environment Program, UNDP, World Bank, and UNCHS, August.

Ankrah, L. (1994), *The Water Supply System in Madina: The Role of the Water Vendor*, B.A. Dissertation, Department of Geography and Resource Development, University of Ghana–Legon.

Armah, N.A. (1993), 'Waste Management', Proceedings of the Ghana Academy of Arts and Sciences, *The Future of Our Cities*, Volume 28, 1989.

Arntzen, J.W. (1994), 'The Contribution of Economic Instruments towards a Sustainable Water Supply in Botswana', in Gieske and Gould, 1994.

Arntzen, J.W., and E.M. Veneendaal (1986), *A Profile of Environment and Development in Botswana*, Institute for Environmental Studies, Free University, Amsterdam, and National Institute of Development Research and Documentation, University of Botswana–Gaborone, October.

Arup Economic Consultants (1991), *Water Use and Affordability*, Final Report to Water Utilities Corporation (WUC), Gaborone, September.

Bahl, R.W., and J.F. Linn (1992), *Urban Public Finance in Developing Countries*, World Bank and Oxford University Press.

Becker, C.M., A.M. Hamer, and A.R. Morrison (1994), *Beyond Urban Bias in Africa: Urbanization in an Era of Structural Adjustment*, Heinemann, James Curry.

Benneh, G., J. Songsore, J.S. Nabila, A.T. Amuzu, K.A. Tutu, Y. Tangtuoru, and G. McGranahan (1993), *Environmental Problems and the Urban Household in the Greater Accra Metropolitan Area (GAMA), Ghana*, Stockholm Environmental Institute.

Botchie, G. (1994), 'Urban Solid Waste in Tema: A Fast Growing Industrial and Port City in Ghana', *African Urban Quarterly*, November.

Braimah, A.A. (1994), *Greater Gaborone Structure Plan (1994-2014)*, Ministry of Local Government, Lands and Housing (MLGLH) and Gaborone City Council (GCC), Draft.

Butcher, C. (1993), 'Urban Low-Income Housing: A Case Study of the Epworth Squatter Settlement Upgrading Programme', Chapter 6 of Zinyama *et al.*, 1993.

Cairncross, S., J.E. Hardoy, and D. Satterthwaite (eds) (1990), *The Poor Die Young: Housing and Health in Third World Cities*, Earthscan Publications.

Central Statistical Office (CSO) (1994), *Census 1992: Provincial Profile, Harare*, June.

Chikwore, E. (1993), 'Harare: Past, Present and Future', Chapter 1 of Zinyama *et al.*, 1993.

City of Gaborone (COG) (1993), *Private Provision of Social Services: Phase II Feasibility Report on Disposal Alternatives*, First Draft, 15 March.

City of Harare (COH) (various years), *Annual Report of the Director of Works*, Department of Works, October.

City of Harare (COH) (various years), *City Treasurer's Report and Accounts*, Harare.

City of Harare (COH) (1993), *Master Plan for Water Distribution*, Volume 3, Interim Master Plan, Director of Works and Nicholas O'Dwyer & Partners, Preliminary Issue, June.

Colclough, C., and S. McCarthy (1980), *The Political Economy of Botswana: A Study of Growth and Distribution*, Oxford University Press.

Colquhoun, B., H. O'Donnell, and Partners (1990a), *Case Study of Urban Infrastructure Operations and Maintenance in Southern Africa Region*, Final Report, Urban Management Program, U.N. Center for Human Settlements, November.

Colquhoun, O'Donnell and Partners (1990b), *Study of Standards and Cost of Infrastructure and Shelter in Botswana*, Final Report, Volume I, November.

Commonwealth Relations Office (various years), *Annual Report of the Bechuanaland Protectorate*, Controller of Stores, Mafeking.

Convard, N.S., and L.J. O'Toole (1993), *Integrated Assessment of Hazardous Waste Management in Botswana*, Water and Sanitation for Health (WASH) Project, USAID, WASH Task No. 472TAS, July.

Coolidge, J.G., R.C. Porter, and Z.J. Zhang, (1993), *Urban Environmental Services in Developing Countries*, EPAT/MUCIA Working Paper No. 9, December.

Cour, J.M. (1985), *Macroeconomic Implications of Urban Growth: Interactions between Cities and Their Hinterlands*, World Bank, September.

Cumming, S.D. (1993), 'Post-Colonial Urban Residential Change in Harare: A Case Study', Chapter 13 of Zinyama *et al.*, 1993.

Darkwa, S.N. (1990), *A Study on the Causes of Poor Sanitary Conditions in the Residential Areas in the Municipality of Accra — A Case Study of Labadi*, B.A. Dissertation, Department of Geography and Resource Development, University of Ghana — Legon.

Dasgupta, P. (1993), *An Inquiry into Well-Being and Destitution*, Oxford University Press.

de Kruijff, G.J. (1981), *Infrastructure Design Standards in Zimbabwe*, Housing Development Services Branch, Ministry of Local Government and Housing, 13 April.

Dijeng, G.M. (1994), 'Private Sector Participation in Collection of Solid Waste', Prepared for Wastecon '94 Conference, Somerset West, South Africa, 27-29 September.

Dintwa, B. (ed) (1984), *The Eleventh National District Development Conference*, Ministry of Local Government and Lands (now MLGLH), Gaborone, 9-16 November.

Dohrman, J.A., D.R. Manning, and B. Spielmann (1991), *A Proposal for the Improvement of Solid Waste Collection and Disposal for the City of Gaborone*, International City Managers Association, Washington, D.C. June.

Dorkenoo, T. (1995), 'The Rains, Drainage and Floods in Accra — Is the Past Guiding Us?' [Accra] *Weekly Spectator*, (Part 1) 18 March and (Part 2) 25 March.

Engmann, E.Y.S. (1990), *Urban Management Program: Case Study on Infrastructure Operations and Maintenance*, U.N. Center for Human Settlement (Habitat), May.

Feddema, J.P. (1977), *Housing for the Poor in Gaborone, the Young Capital of Botswana*, Free University, Amsterdam, February.

Fischer, C., and R.C. Porter (1993), 'Different Environmental Services for Different Income Groups in LDC Cities: Second-Best Efficiency Arguments', Department of Economics Working Paper No. 93-28, The University of Michigan, November.

136

Gaborone City Council (GCC) (various years), *Audit Report and Accounts*, Gaborone.

Ghana Water and Sewerage Corporation (GWSC) (1993), *Corporate Plan, 1994-1997*.

Gibb & Partners (1983), *Gaborone Sewerage Study Master Plan*, Ministry of Local Government and Lands (now MLGLH), December.

Gibb & Partners (1988), *Gaborone Water Supply Master Plan: Pre-Investment Study*, Water Utilities Corporation, July.

Gibb & Partners (1989), *Gaborone Sewerage Study Master Plan: 1988 Up-Date*, Ministry of Local Government and Lands (now MLGLH), January.

Gieske, A., and J.E. Gould (eds) (1994), *Integrated Water Resources Management Workshop, 1994*, Departments of Environmental Science and Geology, University of Botswana, 17-18 March.

Gould, J.E. (1984), *Rainwater Catchment — Possibilities for Botswana*, Botswana Technology Centre, Technical Paper No. 1, April.

Gould, J.E. (1994), 'Long-Term Water Resource Management in Botswana: The Case for Controlling Demand', in Gieske and Gould, 1994.

Gould, J.E., M.A. Humme, and R.L. Senior (eds) (1995), *Workshop on 'Waste' Recycling*, Organized by Somarelang Tikologo (Environment Watch Botswana), 10 May.

Government of Botswana (GOB) (1991a), *Botswana National Water Master Plan, Final Report: Volume 1 — Summary*, Prepared by: Snowy Mountain Engineering Corp.; WLPU Consultants; and Swedish Geological International AB, July.

Government of Botswana (GOB) (1991b), *Botswana National Water Master Plan, Final Report: Volume 3 — Economics, Demography, and Water Demands*, Prepared by: Snowy Mountain Engineering Corp.; WLPU Consultants; and Swedish Geological International AB, January.

Government of Botswana (GOB) (1991c), *Botswana National Water Master Plan, Final Report: Volume 9 — Sanitation*, Prepared by: Snowy Mountain Engineering Corp.; WLPU Consultants; and Swedish Geological International AB, March.

Government of Botswana (GOB) (1991d), *Botswana National Water Master Plan, Final Report: Volume 10A — Water Development Strategies*, Prepared by: Snowy Mountain Engineering Corp.; WLPU Consultants; and Swedish Geological International AB, June.

Government of Ghana (GOG) (1988), *Urban Utilities and Municipal Services*, Final Report, Ministry of Local Government, Town and Country Planning Department, U.N. Center for Human Settlements (Habitat), November.

Government of Ghana (GOG) (1992a), *Accra Residential and Market Upgrading Study: Feasibility Report*, Ministry of Works and Housing,

Technical Services Centre, and Bidex Consult, Volume 1: General Report and Proposed Interventions, May.

Government of Ghana (GOG) (1992b), *Ghana Housing Data: Report of Housing Statistics*, Policy Planning and Evaluation Unit (PPEU), Ministry of Works and Housing, December.

Government of Ghana (GOG) (1992c), *Strategic Plan for the Greater Accra Metropolitan Area*, Volume 1 - Context Report, Accra Planning and Development Program, December.

Government of Ghana (GOG) (1992d), *Strategic Plan for the Greater Accra Metropolitan Area*, Volume 2 - Strategic Plan, Accra Planning and Development Program, December.

Government of Ghana (GOG) (1993), *A National Strategy for Sustainable Human Settlements in Ghana*, The National Task Force, Final Report, June.

Government of Ghana (GOG) (1994), *An Environmental Profile of Accra Metropolitan Area*, Accra Metropolitan Assembly, Ministry of Local Government, UN Center for Human Settlements (Habitat), E.Y.S. Engmann, and A.T. Amuzu, July.

Government of Ghana (GOG) (1995), *Budget Statement and Economic Policy for the Financial Year, 1995*, April.

Gugler, J., and W.G. Flanagan (1978), *Urbanization and Social Change in West Africa*, Cambridge University Press.

Hardoy, J.E., D. Mitlin, and D. Satterthwaite (1992), *Environmental Problems in Third World Cities*, Earthscan.

Harvey, C., and S.R. Lewis, Jr. (1990), *Policy Choice and Development Performance in Botswana*, Macmillan.

The Herald (1995a), 'Harare's Water Supply Dams Critically Low', Harare, Zimbabwe, 20 March.

The Herald (1995b), 'Shacks Built for Renting in Epworth', Harare, Zimbabwe, 21 March.

The Herald (1995c), 'Less Talk, More Action on City Water Crisis', Harare, Zimbabwe, 21 March.

The Herald (1995d), 'Harare Threatens to Impose Water Rationing', Harare, Zimbabwe, 24 March.

The Herald (1995e), 'Water Rationing in Harare Imminent', Harare, Zimbabwe, 1 April.

The Herald (1995f), 'Strict Water Measures Announced', Harare, Zimbabwe, 19 October.

HIFAB International AS (1987), *Sanitation Sector Management Study*. Ministry of Local Government and Lands (now MLGLH) and the World Bank, March.

Huq, M.M. (1989), *The Economy of Ghana: The First 25 Years since Independence*, Macmillan Press.

Khupe, J.S.N. (1995), 'Water Supply, Sewerage and Waste Management for Gaborone', Presented at the Royal Swedish Academy of Sciences Seminars on the Sustainable City, 22-24 January.

Letsholo, J.M.O. (1982), 'The New Towns of Botswana', in R.R. Hitchcock and M.R. Smith (eds), *Settlement in Botswana*, Heinemann.

Linn, J.F. (1983), *Cities in the Developing World*, Oxford University Press.

Maendeleo (Botswana) (1992), *Review of the Self-Help Housing Agency*. Ministry of Local Government, Lands and Housing (MLGLH), May.

Mafico, C.J.C. (1991), *Urban Low Income Housing in Zimbabwe*, Avebury.

Mason, J.P. (1979), *Social Research of Resident Preference, Need and Ability to Pay*, USAID, September.

Masundire, H.M. (1994), 'Waste Water Re-Use and Cultural Eutrophication: A Case Study', in A. Gieske and J. Gould (eds), *Integrated Water Resources Management Workshop, 1994*, Departments of Environmental Science and Geology, University of Botswana, 17-18 March.

Mbizi, C. (1990), *Upgrading Projects as a Viable Alternative Squatter Problem Solution: A Focus on Epworth*, B.Sc. (Hons) Dissertation, Department of Rural and Urban Planning, University of Zimbabwe.

Midweek Sun (1995), 'Collect-a-Can Success Story', Gaborone, 12 April.

Ministry of Lands, Agriculture, and Water Development (MLAWD) (1993), *Harare Water Supply Study, Raw Water Sources*, Prefeasibility Report, Department of Water Development, September.

Ministry of Local Government, Lands and Housing (MLGLH) (1992), *Urban Development Standards, 1992*, May.

Molebatsi, C.O. (1995), 'The Sustainable City: Gaborone — General Information', presented at the Royal Swedish Academy of Sciences Seminars on the Sustainable City, 22-24 January.

Morgan, P. (1987), 'A Case Study in Epworth Zimbabwe', J. Pickford (ed), *Developing World Water*, Grosvenor Press International.

Mosha, A.C. (1992), 'The Design and Architecture of the Built Environment in Botswana', presented at an International Workshop on Creative Design and Architecture for a Better Living, University of Botswana, 6-10 June.

Mosha, A.C. (1995), 'The Planning and Management of the City of Gaborone', presented at the Royal Swedish Academy of Sciences Seminars on the Sustainable City, 22-24 January.

Mosienyane, L.L. (1995), 'SHHA – The Botswana Self Help Housing Agency, A Success Story, But Is It Sustainable?' presented at the Royal Swedish Academy of Sciences Seminars on the Sustainable City, 22-24 January.

Moyo, S., P. O'Keefe, and M. Sill (1993), *The Southern African Environment: Profiles of the SADC Countries*, Earthscan Publications.

Mubvami, T., and T. Korsaeth (1995), *Geographical Information System for Technical Infrastructure Management in Urban Councils*, Interconsult International A/S, 7 February.

Oddoye, L. (1985), *The Pollution Problem in Accra: A Case Study of Both the Domestic and Industrial Sectors*, B.A. Dissertation, Department of Geography and Resource Development, University of Ghana — Legon.

Olsen, R.Y. (1989), Impact of Design Standards on Costs of Urban Development in Botswana, PADCO, December.

Otoo, S.N. (1993), 'Health Aspect of Water and Waste Management', in Proceedings of the Ghana Academy of Arts and Sciences, *The Future of Our Cities*, Volume 28.

Overseas Development Administration (ODA) (1993), *Waste Disposal in Kumasi, Ghana: Pilot Project*, Final Report, Volume 1, October.

Patel, D.H. (1988), 'Government Policy and Squatter Settlements in Harare, Zimbabwe', Chapter 14 of R.A. Obudho and C.C. Mhlanga (eds), *Slum and Squatter Settlements in Sub-Saharan Africa*, Praeger.

Pokoo, J.E. (1994), *Problems of Waste Disposal in Bubuashie*, B.A. Dissertation, Department of Geography and Resource Development, University of Ghana — Legon.

Porter, R.C. (1978a), 'A Model of the Southern-African-Type Economy', *American Economic Review*, December.

Porter, R.C. (1978b), 'A Social Benefit-Cost Analysis of Mandatory Deposits on Beverage Containers', *Journal of Environmental Economics and Management*, December.

Porter, R.C. (1983), 'Michigan's Experience with Mandatory Deposits on Beverage Containers', *Land Economics*, May.

Porter, R.C. (1984), 'Apartheid, the Job Ladder, and the Evolutionary Hypothesis: Empirical Evidence from South African Manufacturing, 1960-77', *Economic Development and Cultural Change*, October.

Porter, R.C. (1993a), *Providing Urban Environmental Services in Developing Countries*, EPAT/MUCIA Policy Brief No. 4, November.

Porter, R.C. (1993b), 'Toward an Economic Theory of the Apartheid City', in L. Stetting, K.E. Svendsen, and W.E. Yndgaard (eds), *Global Change and Transformation*, Copenhagen Studies in Economics and Management No. 1, Handelshojskolens Forlag.

Porter, R.C. (1996), *The Economics of Water and Waste: A Case Study of Jakarta, Indonesia*, Avebury Press. (An earlier draft is available as EPAT/MUCIA Supplementary Paper No. 1, May 1995.)

Potts, D. (1994), 'Urban Environmental Controls and Low-Income Housing in Southern Africa', Chapter 12 of H. Main and S. W. Williams (eds), *Environment and Housing in Third World Cities*, Wiley.

Preble, R.E. (1984), *Helping Ghana Search for Water*, Water and Sanitation for Health Project (WASH) Field Report No. 132, September.

Rakodi, C., and N.D. Mutizwa-Mangiza (1989), *Housing Policy, Production and Consumption: A Case Study of Harare*, RUP Teaching Paper Number 3, Department of Rural and Urban Planning, University of Zimbabwe, November.

Rakodi, C., and N.D. Mutizwa-Mangiza (1990), 'Housing Policy: Production and Consumption in Harare: A Review', Part I, *Zambezia*, Volume 17, Number 1.

Rambanapasi, C.O. (1994), 'Chitungwiza — The Case Study of a Dormitory Town in Zimbabwe', in Wekwete and Rambanapasi, 1994.

Ray, B.T. (1995), *Environmental Engineering*, PWS Publishing Company.

Republic of Botswana (ROB) (1968), *National Development Plan, 1968-73*, Gaborone, August.

Republic of Botswana (ROB) (1973), *National Development Plan, 1968-73: Part I — Policies and Objectives*, Ministry of Finance and Development Planning, Gaborone, August.

Republic of Botswana (ROB) (1977), *National Development Plan, 1976-81: Part I — Policies and Objectives*, Ministry of Finance and Development Planning, Gaborone, May.

Republic of Botswana (ROB) (1983), *Gaborone Sewerage Study Master Plan*, Ministry of Local Government, Lands, and Housing (MLGLH) and Sir Alexander Gibb & Partners, December.

Republic of Botswana (ROB) (1987), *South East Botswana Water Development Study*, Final Report, Department of Water Affairs and Sir M. MacDonald & Partners, December.

Republic of Botswana (ROB) (1991a), *Magnitude and Sources of Water Pollution in Botswana*, Final Report, Department of Water Affairs and Watermeyer, Legge, Piesold & Uhlmann Consultants, January.

Republic of Botswana (ROB) (1991b), *National Development Plan, 1991-97: Part I — Policies and Objectives*, Ministry of Finance and Development Planning, Gaborone, December.

Republic of Botswana (ROB) (1993), *Feasibility Study for Immediate Re-Use of Effluent*, Volume I, Ministry of Local Government, Lands, and Housing (MLGLH) and Sir Alexander Gibb & Partners, January.

Republic of Ghana (ROG) (1959), *Second Development Plan (1959-1964)*.

Republic of Ghana (ROG) (1964), *Seven-Year Plan for National Reconstruction and Development (1963/64-1969/70)*.

Republic of Ghana (ROG) (1968), *Two-Year Development Plan (1968-1970)*.

Republic of Ghana (ROG) (1977), *Five-Year Development Plan (1975/76-1979/80)*, Part II.

Republic of Ghana Statistical Service (ROGSS) (1989), *Ghana Living Standards Survey*, First Year Report, September 1987-August 1988, August.

Republic of Ghana Statistical Service (ROGSS) (1995), *Ghana Living Standards Survey*, Third Phase Year Report, September 1991-August 1992, April.

Roth, G. (1987), *The Private Provision of Public Services in Developing Countries*, World Bank and Oxford University Press.

Sandy Vorster Market Research (1994), *Returnable Bottle Survey*, 18 May.

Seager, D. (1977), 'The Struggle for Shelter in an Urbanizing World: A Rhodesian Example', *Zambezia*, Volume 5, Number 1.

Segosebe, E.M., and C. van der Post (1991), *Urban Industrial Solid Waste Pollution in Botswana*, National Institute of Development Research and Documentation (NIR), Working Paper No. 57, University of Botswana, April.

Senah, K. (1989), 'Problems of the Health Care Delivery System', in E. Hansen and K.A. Ninsin (eds), *The State, Development and Politics in Ghana*, Codesiria Book Series.

Serathi, E. (1994), 'Urban Development and Planning in Botswana', in K.H. Wekwete and C.O. Rambanapasi (eds), *Planning Urban Economies in Southern and Eastern Africa*, Avebury.

Singh, K.A., and R.C. Porter (1995), *Jakarta, Indonesia: The Economics of Water and Waste*, EPAT/MUCIA Case Study No. 4, November.

Somarelang Tikologo (Environment Watch) (1995), *Newsletter, 1995*, Volume 1, Number 1, March.

Songsore, J. (1992), *Review of Household Environmental Problems in the Accra Metropolitan Area, Ghana*, Working Paper, Stockholm Environment Institute (SEI).

Stephens, C., *et al.* (1994), *Environment and Health in Developing Countries: An Analysis of Intra-Urban Mortality Differentials Using Existing Data*, Collaborative Studies in Accra (Ghana) and Sao Paulo (Brazil), London School of Hygiene and Tropical Medicine, Draft, April.

Tahal Consulting Engineers Ltd. (1980), *Accra-Tema Water Supply and Sewerage Project: Review of the Master Plan*, August.

Tahal Consulting Engineers Ltd. (1987), *Rehabilitation and Emergency Works — ATMA Phase III*, January.

Teedon, P. (1990), 'Contradictions and Dilemmas in the Provision of Low-Income Housing: The Case of Harare', Chapter 12 of P. Amis and P. Lloyd (eds), *Housing Africa's Urban Poor*, Manchester University Press.

Tevera, D.S. (1993), 'Waste Recycling as a Livelihood in the Informal Sector: The Case of Harare's Teviotdale Dump Scavengers', Chapter 8 of Zinyama *et al.*, 1993.

Tevera, D.S. (1994), 'Dump Scavenging in Gaborone, Botswana: Anachronism or Refuge Occupation of the Poor?', *Geografiska Annaler*, 76 B(1).

United Nations Development Program (UNDP) (various years), *Human Development Report*, Oxford University Press.

van Hoffen, P. (1975), 'The City of Salisbury Report on the African Affairs Section of the Urban Plan', *The Rhodesian Journal of Economics*, September.

van Nostrand, J. (1982), *Old Naledi: A Village Becomes a Town*, James Lorimer & Company.

Waste Management Department (WMD) (1993), *Determination of Major Planning Data for Solid Waste Management in the Accra Metropolis*, Accra Metropolitan Authority, December.

Waste Management Department (WMD) (1994a), *Waste Disposal Project: Accra, Ghana*, March.

Waste Management Department (WMD) (1994b), *Public and Private Sector Management of Solid Waste in Accra*, Accra Metropolitan Authority, November.

Waste Management Department (WMD) (1995a), *Management Information System*, Accra Metropolitan Authority, January.

Waste Management Department (WMD) (1995b), *Privatisation of Solid Waste Collection Services in Accra*, Accra Metropolitan Authority, July.

Water and Sanitation for Health (WASH) Project (1986), *Economic and Affordability Analysis of Sanitation Alternatives for Self-Help Housing Areas in Botswana*, WASH Field Report No. 148, January.

Water Utilities Corp (WUC) (various years), *Annual Report and Accounts*, Gaborone.

Wekwete, K.H. (1992), 'Africa', Chapter 5 of R. Stren, R. White, and J. Whitney (eds), *Sustainable Cities: Urbanization and the Environment in International Perspective*, Westview Press.

Wekwete, K.H. (1994), 'Urbanization, Urban Development and Management in Zimbabwe', in Wekwete and Rambanapasi, 1994.

Wekwete, K.H., and C.O. Rambanapasi (eds) (1994), *Planning Urban Economies in Southern and Eastern Africa*, Avebury.

Wilkinson, N.J. (1986), *Water Conservation in Urban Botswana*, Botswana Technology Centre, Technical Paper No. 7, July.

World Bank (various years), *World Development Report*, Oxford University Press.

World Bank (1989a), *Ghana: Urban Sector Review*, Infrastructure Operations Division, Western Africa Department, Africa Region, Report No. 7384-GH, June.

World Bank (1989b), *Staff Appraisal Report: Republic of Ghana Water Sector Rehabilitation Project*, Report No. 7598-GH, Infrastructure Operations Division, Country Department IV, Africa Region, 18 May.

World Bank (1989c), *Zimbabwe: Urban Sector and Regional Development Project*, Infrastructure Operations Division, Southern Africa Department, Report No. 7619-Zim, 8 May.

World Bank (1993a), *Ghana 2000 and Beyond: Setting the Stage for Accelerated Growth and Poverty Reduction*, Report No. 11486-GH, Africa Regional Office, Western Africa Department, February.

World Bank (1993b), *Staff Appraisal Report: Ghana Community Water and Sanitation Project*, Report No. 12406-GH, Infrastructure Operations Division, Country Department IV, Africa Region, 29 December.

World Bank (1994a), *Republic of Ghana Urban Development Strategy Review*, Report No. 13567-GH, Infrastructure Operations Division, Western Africa Department, Africa Region, 22 December.

World Bank (1994b), *Zimbabwe: Urban Development Project*, Infrastructure Operations Division, Southern Africa Department, Report No. 13832.23, December.

World Bank (1994c), *World Development Report, 1994*, Oxford University Press, Washington.

Zindere, S. (1991), *Implications of Low-Income Housing Policies on Affordability: A Focus on Minimum Housing Standards in Glen Norah Extension, Harare*, B.Sc. (Hons) Dissertation, Department of Rural and Urban Planning, University of Zimbabwe.

Zinyama, L.M. (1993), 'The Evolution of the Spatial Structure of Greater Harare: 1890 to 1990', Chapter 2 of Zinyama *et al.*, 1993.

Zinyama, L.M., D.S. Tevera, and S.D. Cumming (eds) (1993), *Harare: The Growth and Problems of the City*, University of Zimbabwe Publications.

Ziracha, M.R. (1989), *Sanitation Standards in Low Cost Housing: A Case Study of Epworth*, B.Sc. Dissertation, Department of Rural and Urban Planning, University of Zimbabwe.

144